W0230555

DEVELOPMENTS IN RUBBER TECHNOLOGY—2
Synthetic Rubbers

THE DEVELOPMENTS SERIES

Developments in many fields of science and technology occur at such a pace that frequently there is a long delay before information about them becomes available and usually it is inconveniently scattered among several journals.

Developments Series books overcome these disadvantages by bringing together within one cover papers dealing with the latest trends and developments in a specific field of study and publishing them within *six months* of their being written.

Many subjects are covered by the series, including food science and technology, polymer science, civil and public health engineering, pressure vessels, composite materials, concrete, building science, petroleum technology, geology, etc.

Information on other titles in the series will gladly be sent on application to the publisher.

DEVELOPMENTS IN RUBBER TECHNOLOGY—2
Synthetic Rubbers

Edited by

A. WHELAN and K. S. LEE

National College of Rubber Technology,
Holloway, London, UK

APPLIED SCIENCE PUBLISHERS LTD
LONDON

APPLIED SCIENCE PUBLISHERS LTD
RIPPLE ROAD, BARKING, ESSEX, ENGLAND

British Library Cataloguing in Publication Data

Developments in rubber technology.—(Developments
series).
2: Synthetic rubbers
1. Elastomers   2. Rubber
I. Whelan, Anthony   II. Lee, K. S.
III. Series
678   TS1925

ISBN-13: 978-94-009-8110-2   e-ISBN-13: 978-94-009-8108-9
DOI: 10.1007/978-94-009-8108-9

WITH 80 TABLES AND 57 ILLUSTRATIONS

© APPLIED SCIENCE PUBLISHERS LTD 1981
Softcover reprint of the hardcover 1st edition 1981

All rights reserved. No part of this publication may be reproduced, stored in
a retrieval system, or transmitted in any form or by any means, electronic,
mechanical, photocopying, recording, or otherwise, without the prior
written permission of the publishers, Applied Science Publishers Ltd,
Ripple Road, Barking, Essex, England

PREFACE

This book is intended for those people who have a knowledge or understanding of rubber materials and processes but who wish to update their knowledge. It should be read in conjunction with *Developments in Rubber Technology—1* as that volume discussed developments in natural rubber and selected special purpose synthetic rubbers as well as additives.

The authors have been selected for their expertise in each particular field and we, as editors, would like to express our appreciation to the individual authors and also to their companies. Such a book would be impossible to produce without such active cooperation as we have received.

Volumes 1 and 2 of *Developments in Rubber Technology* cover rubbers which are processed and vulcanised in the traditional manner. It is appreciated that the omission of non-vulcanised rubber materials (the so-called thermoplastic elastomers) will be unwelcome to many readers but it is intended, because of the size of the subject, to cover these materials in a subsequent volume.

A.W.
K.S.L.

4

CONTENTS

LIST OF CONTRIBUTORS

J. C. BAMENT

Elastomers Research Laboratory, Du Pont (UK) Ltd, Maylands Avenue, Hemel Hempstead, Herts HP2 7DP, UK

H. H. BERTRAM

Bayer AG, D-5090 Leverkusen, Bayerwerk, West Germany

J. A. BRYDSON

National College of Rubber Technology, The Polytechnic of North London, Holloway, London N7 8DB, UK

L. CORBELLI

Montedison S.p.A., DIMP/CER, Piazzale Privato Donegani 14, 44100 Ferrara, Italy

R. J. CUSH

Dow Corning Ltd, Barry, Glamorgan CF6 7YL, UK

W. D. GUNTER

Polysar Ltd, Sarnia, Ontario, Canada N7T 7M2

J. G. PILLOW

Elastomers Research Laboratory, Du Pont (UK) Ltd, Maylands Avenue, Hemel Hempstead, Herts HP2 7DP, UK

M. J. SHUTTLEWORTH

Compagnie Française Goodyear, Centre Technique, Avenue des Tropiques, 91941 Les Ulis, France

A. A. WATSON

Compagnie Française Goodyear, Centre Technique, Avenue des Tropiques, 91941 Les Ulis, France

H. W. WINNAN

Dow Corning Ltd, Barry, Glamorgan CF6 7YL, UK

Chapter 1

TRENDS IN THE USAGE OF RUBBERY MATERIALS

J. A. BRYDSON

National College of Rubber Technology,
The Polytechnic of North London, Holloway, London, UK

SUMMARY

The future pattern of rubber usage is predicted in the light of expected global changes in living standards and, specifically, methods of transportation. It is concluded that the car and the lorry will become increasingly favoured methods of transport. Whilst there will continue to be an increasing demand for car ownership the increasing cost of energy is likely to lead to smaller, lighter cars as well as more careful patterns of usage. This will lead to a reduction in the amount of rubber used per car. Whilst non-automotive applications of rubber will continue to be developed the use of rubber will largely be determined by the automotive industries. This growth of rubber consumption will require an increase in car ownership to more than offset the amount of rubber used per car.

The outlook for individual types of rubber, which are far from uniform, are reviewed. It is concluded that the rubber industry will have to work much harder and more effectively than formerly to survive the next decade, and that companies that fail to undertake the necessary development work to ensure that their products find a ready market and those that make inefficient use of manpower, equipment and materials cannot be expected to enjoy a long-term prosperity.

1. INTRODUCTION

Those of us who attempt to predict future trends cannot win. If our predictions turn out to be wrong, then we are simply discredited. If, on the

1

other hand, our predictions turn out to be correct, then we are likely to be told that what we had predicted was obvious in any case. A less pleasant alternative is that we may make predictions which, when made, were not what people wanted to hear and when they prove to be correct, the person making the prediction is, somehow, held to blame for the occurrence of the event.

Whilst predictions from an individual's point of view are games for masochists, there is a real need that they should be made. No organisation which hopes to have a successful, long-term future could reasonably expect to exist without making some assessment of what is likely to happen in the future. What, however, must be recognised is that situations may be expected to change and, consequently, it is necessary to update predictions as these changes occur. It is therefore reasonable that between writing this chapter and its publication, important changes may have occurred and so my predictions may need to be modified. It will, however, be my aim in preparing this chapter to explain the basis for the predictions so that where changes have occurred in the underlying situation, the reader will be able to modify and adapt my predictions appropriately.

In order to try and predict trends in the usage of a particular material, it is necessary to consider a variety of underlying trends. Some of these underlying trends may be very long-term and others very short. It is clearly most sensible to determine the primary long-term trends and then to superimpose on them the secondary trends of shorter duration.

In the specific case of the consideration of trends in the usage of rubbery materials, the following global underlying trends may be identified:

1. Changes in standards of living
2. Changes in patterns of transportation
3. Changes in the design of automotive equipment
4. Changes in tyre design and the use of rubber in other automotive applications
5. Changes in non-automotive uses for rubber
6. Changes in raw material supplies
7. Trends in legislation, for example in areas associated with health, safety and toxicity

2.   LONG-TERM UNDERLYING TRENDS

The first two of the above list of underlying trends are obviously of a very long-term nature. The 20th century has, with the two interludes of the two world wars, seen a period of progressive increase in the global standard of living. In the first quarter of this century, the bulk of the improvement in standards of living occurred in Europe, in North America and among the white population of South Africa. Even in these parts of the world, many of the improvements were confined to a fairly small and fortunate part of the population. It is only since the Second World War that such items as cars, telephones and refrigerators have become available to a European on an average national wage. During the second half of this century, we have seen substantial increases in the standard of living of other countries. Expectations are increasing and there now seems no *a priori* reason why a large part of the population of the so called Third World cannot also begin to enjoy such a higher standard of living.

Without doubt, the major change that has occurred during this century has been the great advance in communication. This may be taken to include developments in transportation and in particular in the development of the motor car and the aeroplane. Since the automotive industry is far and away the major consumer of rubbery materials, it behoves us to try and discern trends in transportation. In spite of all the criticisms levelled against it, there is little doubt, that as far as the individual is concerned, the motor car is a wonderful asset. It gives him considerable flexibility and independence when travelling as well as apparently being able to satisfy all sorts of psychological needs. Car ownership is a habit which can only be broken with the greatest of difficulty. Thus it will take exceptional pressures for there to be a reduction in car ownership and use.

Much has been said about the energy crisis and the huge increases in the price of petrol that have resulted from this. It has to be said that to date, at least in the UK, these increases are more apparent than real. Table 1 shows how the price of a gallon of petrol and of a car have changed over the past 25 years in comparison with other common everyday costs. Clearly in real terms, petrol is cheaper now than it was a quarter of a century ago.

We are, of course, in a dynamic situation. The price of a barrel of crude oil has risen about 1500% in the seven years between 1973 and 1980. It is reasonable to expect that this will continue to rise for some time until other sources of energy become competitive. A stage will be reached when the oil shales become a commercial source of fuel, when the harnessing of light energy becomes a feasible possibility and when wave power also becomes worth tapping. Furthermore, one might expect sugar to become a source of fuel and that it may be possible to develop cars which run on hydrogen. There seems very little evidence to date that, up to the time that this occurs, car owners will give up their cars, although they may well be more prudent

TABLE 1

SOME PRICE INCREASES IN THE LONDON AREA BETWEEN 1955 AND 1980

	Price (p)		Price ratio
	1955	1980	
First class letter	1	12	12
Evening newspaper	0·4	12	30
Suburban cinema seat	10	150	15
Typical underground railway fare[a]	3	80	27
Daily water rate (typical suburban semi-detached house)	1	10	10
Small family car[b]	$6·5 \times 10^4$	32×10^4	5
Petrol (1 gal)	23	128	5·6

[a] Southgate–Holloway Road.
[b] 1955 Morris Minor 1000; 1980 Austin Allegro.

in their use. On the other hand, one may anticipate a considerable increase in car ownership around the world.

As cars appear to continue to be the increasingly preferred manner of personal transportation so the truck appears likely to dominate the transportation of goods more and more. Large areas of the world are remote from the railway and as standards of living improve, particularly in remote parts, there will be increasing goods traffic. Thus, although railways might recover some of the market in Europe where there is a dense railway network, the overall picture indicates increasing use of the truck. The day of the automobile seems far from over.

3.   CHANGES IN CAR AND CAR TYRE DESIGN

Whilst cars are likely to continue to be the preferred means of personal transport both national governments and individuals are very conscious of costs, particularly fuel costs. This has led to the increasing popularity of vehicles which are more economical in terms of fuel consumption. In some places, such as in the United States, this trend is being backed by legislation. Thus cars are becoming lighter. This helps to reduce tyre wear and also means smaller tyres. These factors indicate a reduction in rubber usage although the life of a small tyre is usually less than that of a large one, everything else being equal. One prediction is that between 1980 and 1990 the average weight of an American car will drop from 2816 to 2522 lb including a drop in rubber usage from 144 to 128 lb. If, however, the smaller tyre had a life 10 % less than that of a larger tyre, on the same size car the usage of rubber, taking into account replacement needs, would be substantially unchanged. The rubber industry is still adjusting itself to the worldwide acceptance of the radial tyre. This came much later in North America than in Europe and the repercussions of the advent of tyres of much greater longevity in this major rubber market has had an important influence on overall rubber usage. The different requirements of the radial tyre *vis-à-vis* the cross-ply have simultaneously increased the demand for natural rubber relative to SBR.

There has also been a change in the height/width ratio of tyre cross-sections. Where this is brought about by reducing the height and keeping section width constant there is both a saving in the use of rubber in tyre manufacture as well as improved wear resistance—albeit with some loss of road grip. A further trend is the use of higher tyre pressures to reduce rolling resistance—in this case at some loss to ride comfort although this is being overcome by modifications to side wall shape. Overall however, these more recent developments tend to lead to a reduction in rubber usage.

We may conclude this section by saying that whilst there appears to be a trend to increased car ownership, economical driving habits coupled with changes in car and car tyre design are likely to reduce the rate of rubber consumption per car owner.

4. NON-TYRE APPLICATIONS

Tyres remain the major reason for rubber usage. This is clearly shown in Table 2[4] which shows the rubber consumption in various countries in 1968 and 1978. Although there are national variations the overall figure for tyres and tyre products has remained about 60 % of the total. It is also interesting to note that natural rubber has a smaller part of the non-tyre market (19 %) than in the tyre sector (29 %).

Table 3[2] is an estimate, for 1978, of the relative importance of the main end-uses of rubber in western Europe.

It is not surprising that these end-uses are dominated by the general rubber goods which includes parts for cars and other means of transport, for washing machines and other domestic equipment and a wide range of industrial mouldings.

The residual usage of rubber may be divided into two categories: Non-tyre automotive uses and non-automotive uses. It is reasonable to predict

TABLE 2

USE OF NATURAL AND SYNTHETIC RUBBERS IN TYRE AND NON-TYRE PRODUCTS IN MAJOR
PRODUCING COUNTRIES (1968 AND 1978) (THOUSANDS OF TONNES)

	Tyre products				Non-tyre products			
	1968		1978		1968		1978	
	NR^a	SR^b	NR	SR	NR	SR	NR	SR
USA[c]	402	1224	623	1464	189	697	181	1017
UK	93	130	84	118	101	114	55	195
France	86	110	132	178	42	86	31	118
Germany	80	140	117	161	90	113	68	268
Italy	56	76	65	101	44	84	48	164
Japan	125	175	251	415	130	172	103	325
Brazil	28	43	56	125	10	28	16	96
Total of above countries	870	1908	1328	2562	606	1294	502	2183
Grand total	2778		3890		1900		2685	

[a] NR = Natural rubber.
[b] SR = Synthetic rubber.
[c] USA data for 1977.

continuing increase of use of rubber in cars for seals, grommets, springs. mountings, hose and so on. The demands for increased safety and lightness in weight will help rubbers to maintain their areas of application.[3] Of rather more concern is the reliability of some rubber parts. As one who has suffered enormous repair bills due to the failure of rubber components, which were themselves of little cost, there must be concern in this area of use. For over 20 years I drove a particular make of car which I generally found excellent and reliable except for problems consequent on the continual failure of various rubber components. For this reason I changed my brand loyalties. It is reasonable to hope that good rubber technology, stimulated by more stringent specifications by the car manufacturers, can overcome these problems. If this is not done both the market for the car manufacturers and their suppliers will be badly affected.

Outside of the automobile industry usage of rubber has suffered many changes in the past quarter of a century. Applications such as cables, battery boxes, belting, footwear, rainwear and flooring which were entrenched markets for rubber before the Second World War have disappeared with the development of thermoplastic materials. Such uses

TABLE 3
ESTIMATED RELATIVE IMPORTANCE OF MAIN END-
USES OF RUBBER IN WESTERN EUROPE, 1978 (NON-
TYRE APPLICATIONS)

	%
Shoe soles and heels	11
Pipes and tubes	10
Rubber solutions and dispersions	7
Carpeting	6
Cellular rubbers	3
Ebonite	2
Belting	2
Reinforced fabrics	2
Medical goods	1
General rubber goods	56

did not generally demand the property of rubber-like elasticity and the transition may now be considered to be virtually complete.

In its place we have seen increasing use of rubber in aircraft and as an engineering material. The engineering applications include such uses as damping pads, for vibration insulation, rubber springs, dock fenders, bridge bearings and so on where the high elasticity of rubber may be properly utilised. The development of the aerospace industry, in particular of aircraft, has also led to many new uses for rubbery materials. In other words rubbers are now being used where rubbery properties are required and it is reasonable to expect that they will be able to retain most of their current areas of application. If there is to be a global increase in the standards of living then we may expect that the total usage of rubber in non-automotive applications will increase.

5. PROSPECTS FOR INDIVIDUAL GENERAL PURPOSE RUBBERS

It is most unlikely that all rubbers will experience similar usage trends. It is therefore useful to consider the situation with both natural rubber and the synthetic rubbers individually.

5.1. Natural Rubber
In many ways the current position for natural rubber *vis-à-vis* the general

purpose synthetic rubbers is most enviable. During the period 1945–1974 the demand for rubber (both natural plus synthetic) grew at a rate of about 7 % a year. During this time natural rubber supply grew at about 2–3 % a year so that the percentage of the market taken by natural rubber decreased progressively reaching a figure of about 30 % by the early 1970s. At this time about 14 % of the rubber used in North American car tyres was natural rubber with a figure of about 27 % in truck tyres. With the acceptance of radial tyres the demand for natural rubber, with its superior dynamic properties and green strength, has raised these figures to about 25 % and 62 % respectively. It must, however, be pointed out that although the percentage of natural rubber in tyres has increased in this market, this benefit is virtually cancelled out by the longer life of the radial tyre. In other words the use of natural rubber per tyre mile is virtually unchanged. The view has been expressed that at the present time the supply–demand position for natural rubber is about balanced.[4] There does, however, seem reason to believe that if there is an increase in car ownership then there will be an increase in the number of tyres produced per year and there could be some degree of shortage of natural rubber. This may be expected to push the price up to a point where the benefits of using natural rubber are largely negated but this would mean a substantial premium price for natural rubber over the general purpose synthetics. Whilst it is difficult to increase the supply of natural rubber sharply over a short period it is, on the other hand, possible to contract sharply the supply of natural rubber should it become obvious that alternative crops such as palm oil will give much greater returns than natural rubber to the planters.

In an attempt to achieve some sort of independence from both oil-based rubbers and from the tropic-grown hevea rubbers, the American Government has recently announced a plan to develop guayule rubber. This rubber is obtained from a small shrub which grows in the south-west of the USA and in Mexico. Methods of extracting latex are somewhat cumbersome but it is claimed that vulcanisates from this rubber show properties similar to those obtained from hevea.

In summary, it would seem that the growth in supply of natural rubber will be somewhat modest over the next few years but that the demand for this polymer will ensure that it has a premium in price over its general purpose synthetic counterparts.

5.2. Styrene–Butadiene Rubber (SBR)

In contrast to the natural material the prospects for SBR are less satisfactory. The worldwide acceptance of the radial tyre has cut down rubber usage per car, a greater percentage of that usage is going to natural rubber and the situation is aggravated further by the advent of lighter cars and the acceptance of higher tyre pressures. On top of that several new SBR plants have come on-stream in recent years so that at the present time supply capacity exceeds demand. This has led to a depressed price for SBR. At least this severely depressed uneconomic price for SBR has provided some relief for the equally sorely pressed tyre manufacturers.

Such a depressed economic situation for SBR now discourages technical developments. Most observers seem agreed that the solution polymerised SBRs are or can be superior to the emulsion materials. However, the current profitability discourages investment in new plant so that it may well be many years before these improved materials become the norm.

Of even greater concern for the SBR manufacturers is the threat of USA legislation which could ban the use of SBR because it contains, by modern analytical techniques, measurable quantities of styrene monomer. Whilst at the time of writing, this threat appears unlikely to materialise, it is indicative of the influence that health and safety aspects now have on the chemical industries of the world.

It is reasonable to ask whether it would be possible to provide a material which is an alternative to SBR should the above mentioned legislation ever occur. The answer is definitely 'yes'. For example, it has been found that SBR can, to a large extent, be substituted by high-vinyl polybutadiene, although this would require new manufacturing plant and some changes in rubber processing technology.

In spite of all the problems mentioned above, the likelihood seems to be that in the 1980s there will be some overall increase in usage of SBR. A recent prediction by the International Institute of Synthetic Rubber Producers[5] suggests that over the coming decade there will be a 41·5% growth in SBR consumption, which is similar to the predicted growth of 42% for natural rubber. The big difference would appear to be in the relative levels of profitability for these two materials.

5.3. Synthetic Polyisoprene

Up to the present time, the market for synthetic polyisoprene has been only a small fraction of the total rubber market. The low cost of SBR and the ready availability, up until now, of natural rubber, have rather discouraged developments with this material. It has therefore tended to be used for those applications where the marginal difference in properties between the synthetic material and natural rubber are most significant. It is probably fair to say that in the 1970s the greatest interest in synthetic polyisoprene was in

TABLE 4

CONSUMPTION FORECASTS FOR NEW RUBBER (THOUSANDS OF TONNES)

Rubber	Global (excluding CPEC)				USA				CPEC			
	1978 (Actual)	1984 (Forecasts)	1989	% Change 1978–1989	1979 (Forecast)	1986 (Forecast)	1990 (Forecast)	% Change 1979–1990	1978 (Actual)	1984 (Forecast)	1989 (Forecast)	% Change 1978–1989
Natural rubber	2945	3650	4348	48	840	940	1025	22	761	847	926	21
SBR (solid and latex)	3411	4086	4827	42	1454	1608	1759	21				
Polybutadiene	918	1143	1389	51	431	496	554	29				
Polyisoprene	229	317	396	73	80	118	159	99	2543	3747	4937	94
Ethylene–propylene	299	426	553	85	158	202	245	55				
Polychloroprene	289	347	397	37	126	142	157	25				
Butyl	382	461	539	41	158	176	191	21				
Nitrile (solid and latex)	188	234	282	50	81	93	104	28				
Other synthetic	124	167	212	71	57	80	101	77				
Total synthetic	5840	7181	8595	47	2545	2915	3270	28				
Total (natural plus synthetic)	8785	10831	12943	47	3385	3855	4295	27	3304	4594	5857	77

the Eastern Bloc countries as part of a strategy of self-sufficiency. Whilst development in western countries has been somewhat restricted by the availability of the monomer, synthetic polyisoprene is expected to be in substantially increased demand during the 1980s (Table 4). This may be associated with the increased demand for polyisoprene rubbers, both natural and synthetic, for use in radial tyres which is coupled with the prospect of a shortfall in natural rubber production.

5.4. Polybutadiene
Closely related to the synthetic polyisoprenes are the polybutadiene rubbers. These rubbers have been widely used in tyre manufacture in blends with other rubbers, particularly SBR. At the present time, the prospects for this material in this application are not entirely clear and this could have a large effect on the overall prospects for polybutadiene. This rubber does, however, have another important use. It has, to a large extent, replaced SBR as the toughening rubber in high impact polystyrene. Continued growth in the polystyrene market would be expected to cause continued growth in the use of polybutadiene in this application. As already mentioned, prospects for polybutadiene could be affected by legislation against styrene in SBR. The reason is that high-vinyl polybutadiene appears to offer the most satisfactory alternative to SBR that can be made readily available. At the same time, however, legislation against styrene could well affect the use of polybutadiene in a high-impact polystyrene. At the present time, predictions are for an overall growth in the use of polybutadiene which is about average for the various rubbers.

6. PROSPECTS FOR SPECIAL PURPOSE RUBBERS

The predictions of the International Institute of Synthetic Rubber Producers are for something like a 57 % rise in synthetic rubber manufacture during the 1980s. With most of the general purpose rubbers expecting a somewhat lower rise, it follows that the more specialised rubbers will have an above average predicted rate of growth.

6.1. Ethylene–Propylene
The prediction for the highest growth rate (85 %) is for ethylene–propylene rubbers. When these materials were first introduced about 20 years ago, there were, in some quarters, great expectations that these rubbers would find their place amongst the category of general purpose rubbers,

particularly for use with tyres. In the event, these hopes were not realised. The reasons for this were varied and included the high cost of cure-site monomers, curing problems associated with possible stock mix-ups, and the lack of building tack. On the other hand, these materials clearly show very good resistance to ageing, particularly at moderately elevated temperatures and the properties of these rubbers are becoming progressively more appreciated, in particular for use in the automotive field. More and more rubber users are specifying ethylene–propylene rubbers for use in moulding and extrusion applications. To some extent, this is at the expense of natural rubber and SBR but also, for those applications involving reasonable resistance to heat and light ageing, of the polychloroprene rubbers. Furthermore, there is now a steadily increasing demand for the manufacture of thermoplastic polyolefin elastomers and for elastomer-modified polyolefins using ethylene–propylene rubbers. Since with ethylene–propylene rubbers one is starting off with a much lower current consumption tonnage figure, as compared with the general purpose rubbers, such changes in usage can lead to large increases in percentage consumption.

6.2. Polychloroprene
In contrast to the ethylene–propylene rubbers, the prospects for polychloroprene rubbers are less exciting. There are many instances, particularly in the automotive field, where it has been found possible to replace polychloroprene rubbers, which have been used for good light, oxygen and ozone resistance, by the less expensive ethylene–propylene rubbers. There has also been some substitution of the polychloroprenes by chlorosulphonated polyethylenes such as Hypalon. In part, this has been due to sharply increasing costs for this rubber. Another reason for moderate growth prospects with the polychloroprenes is that of toxicity hazards in their processing and use. One of the most used curing agents, ethylene thiourea, has been associated with a number of health hazards and difficulties in finding a substitute material have encouraged rubber manufacturers to look elsewhere. In addition, there have been some questions raised concerning the toxicity of the monomer.

6.3. Nitrile
The nitrile rubbers are predicted to have approximately average growth. The main applications for this class of rubber are now well established and growth in usage is likely to depend largely on the worldwide economic situation. It is in fact possible that part of the nitrile rubber market may be

lost to some of the other speciality rubbers, particularly as changes in car design are increasing the severity of the demands being made on oil-resistant rubbers. To some extent these have already been countered by the development of new curing systems for nitrile rubbers but these will not provide the answer in every case.

6.4. Butyl
A somewhat similar situation faces the butyl rubbers. These materials are very dependent on the general level of industrial activity and their markets seem to be well established.

7. OPPORTUNITIES FOR SPECIALITY RUBBERS

The materials described in the previous section, although not classified as general purpose rubbers, have a substantial market with similar annual tonnage consumptions. There also exists a wide range of more specialised materials of much lower tonnage consumption for which the term speciality rubber is applied. At the present time world production of these materials is of the order of about 125 000 t and as these include several new materials which have been recently introduced and have yet to realise their full potential it is not surprising that the predictions are for an overall above-average growth for these materials in excess of 70 % during the next decade.

7.1. Acrylic Rubbers
This trend in growth of speciality rubbers is reinforced particularly by new United States regulations. These demand substantial increases in the miles per gallon performance of cars over the next five years which has led to an extensive re-think in car design. Engines are now being designed to run at higher temperatures and this is increasing the potential for some of the speciality rubbers. For example the acrylic rubbers are already showing high growth rates whilst the related ethylene–acrylate rubber (Vamac), although introduced as recently as 1975 by Du Pont, is already finding uses in under-the-bonnet applications such as coolant hoses, spark plug boots and ignition wiring. These rubbers also find use in power steering hose and transmission seals.

7.2. Chlorosulphonated Polyethylenes
Prospects for chlorosulphonated polyethylene are also encouraging. This material was introduced by Du Pont as long ago as 1952. In August 1979

Du Pont announced a 15 % capacity expansion for their material, marketed as Hypalon, to 36 000 t by 1981. The possibility of further production in Europe of this material has also been under study. This material has an enviable reputation in the field of coated fabrics and film sheeting materials and also for wire and cable coverings. More recently there has been interest in unvulcanised materials for roofing and pit lining applications. With substantial changes in price differentials in recent years the chlorosulphonated polyethylenes are now beginning to penetrate some of the markets previously held by polychloroprenes.

7.3. Fluorocarbons
Another well-established group of speciality materials are the fluorocarbon rubbers which are used in applications requiring oil and chemical resistance over a wide temperature range. To date the main uses of these materials have been for O-rings, packings and gaskets. Recent and anticipated legislation in the field of automotive and industrial emission control systems has led to increasing interest in these materials because of their good resistance to hot corrosive gases. Other applications for these rubbers include roll and conveyor belt covers for hot materials, oil and chemical valve linings, pump impellers and flue duct expansion joints. Whilst several types of fluorocarbon rubber have been marketed over the years, at present the bulk of the commercial market is for the vinylidene fluoride–hexafluoropropylene rubbers such as the Vitons supplied by Du Pont. In the mid-1970s Du Pont introduced a new sub-class of fluorocarbon rubbers in which there were no C—H bonds. These materials have exceptional thermal stability but do require highly specialised processing methods. In general the fluorocarbon rubbers are expensive and the newer materials, marketed under the trade name Kalrez, particularly so.

7.4. Silicones
The silicone rubbers show not only very good heat resistance, like the fluorocarbons, but also very good properties at low temperatures. The uses of these materials are well established where operation over a wide service temperature range is required. In recent years, however, these materials have been facing competition, for some of the less stringent applications, from some of the less expensive speciality materials. These include the acrylic rubbers and the ethylene–acrylate rubbers already mentioned. On the other hand there has become available a special group of silicone materials which may be vulcanised at room temperature, the so-called RTV rubbers.

7.5. Other Speciality Rubbers

There also exists a number of speciality rubbers whose applications are either very limited or well established. These include the chlorinated polyethylenes, the epichlorhydrin rubbers, the fluorosilicones, the ethylene–vinyl acetate rubbers, the polysulphides, the nitroso rubbers and the phosphonitrile fluoroelastomers. Changes in the uses of these materials are likely to have very little impact on the usage of general purpose and major special purpose rubbers. They have also been the subject of a recent review.[6]

There is, however, one further speciality rubber which perhaps should be singled out at this stage. In 1975 the French company CdF Chemie announced the availability of polynorbornene elastomers under the trade name of Norsorex. These materials were quite different from any other rubbers prepared hitherto in that the basic polymer is not itself a rubber. It is a thermoplastic powdery product with a glass transition temperature (T_g) of about +35 °C. However, it has a very high compatibility with a large number of aromatic and naphthenic oils and the blends of the polymer with these oils can give rubbers with a very wide range of softness and which may also remain rubbery down to temperatures as low as −60 °C.

Particularly noteworthy of these materials is the high tensile strength which can be achieved with products of very low hardness. For example, it has been claimed that a rubber with a Shore A hardness of 18 has a tensile strength as high as 10 MPa. In addition it is possible to formulate these rubbers with good damping properties over a wide temperature range. These rubbers are already finding applications in vibration insulation and isolation and for highly filled sound deadening parts for the automotive industry. Whilst it is unlikely that these materials will achieve a high production level it is reasonable to expect that they will, to some extent, replace some of the far more widely used rubbery materials.

8.   THERMOPLASTIC ELASTOMERS

One reason for the success of the thermoplastic materials over the rubbers since the Second World War has been that most thermoplastics are capable of being reprocessed without undue difficulty. On the contrary attempts to re-claim vulcanised rubbers fail to give products of the quality of the original unvulcanised material and the cost of re-claiming cannot often be justified by the returns. It is therefore clearly attractive to attempt to

produce rubbers which may be processed like thermoplastics and which do
not need a special vulcanisation process.

8.1. Types of Thermoplastic Elastomer

Over the years a number of approaches have been made with varying
degrees of success. At the present time the following types of thermoplastic
elastomer may be recognised:

1.   Butadiene–styrene block copolymers
2.   Thermoplastic polyurethanes
3.   Polyether–polyester thermoplastic elastomers
4.   Thermoplastic polyolefin elastomers

8.2. Butadiene–Styrene Materials

The butadiene–styrene block copolymers have now been available for over
15 years. Marketed by Shell (Cariflex TR), Phillips (Solprene T) and ANIC
(Europrene), these materials have established markets in the field of shoe
soling, modification of polystyrene and polyolefins, as bitumen additives
and in adhesive formulations. Whilst the shoe soling applications are, to
some extent, subject to the whims of fashion, steady increase in use of these
materials may be expected. Similar comments may be made concerning the
thermoplastic polyurethanes which have also been available for a number
of years. These materials combine the properties of strength, oil resistance
and processibility.

8.3. Polyether–Polyester Rubbers

The polyether–polyester thermoplastic rubbers were first offered by Du
Pont in 1972 under the trade name of Hytrel. More recently, a competitive
material has become available from the Dutch company AKZO under the
trade name Arnitel. These rubbers have a much wider service temperature
range than shown by the SBS materials. They also have good oil and solvent
resistance, good flex fatigue, high resilience, good resistance to mechanical
abuse and other features desirable in a rubber for use as an engineering
material. Several grades are now available ranging in hardness from that of
a hard grade of rubber to something approaching that of a polyolefin. In
spite of their high cost, these materials are finding application in such areas
as hose liners, seals and industrial mouldings.

8.4. Thermoplastic Polyolefins

The SBS polymers, the thermoplastic polyurethanes, and the polyester–

polyether materials may all be considered as somewhat special purpose polymers. Of far more general applicability are the thermoplastic polyolefin rubbers which came into prominence in the late 1970s.[7] These materials are essentially blends of ethylene–propylene rubber and polypropylene. As with the polyester–polyether polymers, they are available in a range of hardness and stiffness from that of a rubber compound to that of polypropylene. Whilst these materials do not have the good oil and heat resistance of the polyester–polyether polymers, they are moderately priced and easily processed on conventional thermoplastics machinery.

They are finding extensive use in the car industry, for example, for car bumpers and radiator grills. They are also of interest for such diverse applications as cable sheathing and ski boots. There is reason to believe that blends of polypropylene with ethylene–propylene rubbers have considerable scope for further development. Whilst it provides the polymer manufacturers with an opportunity to use up spare capacity for making both polypropylene and the ethylene–propylene rubbers the rubber product manufacturers are less likely to gain much from their existence. This is because most of the processing being carried out with these materials is done by thermoplastics converters rather than the rubber industry.

9. VARIATIONS IN MARKETS

In the preceding sections predictions of wide variations in growth rates have been made for differing rubbery polymers. It is little more than a statement of the obvious that there will be wide variations in growth and success rates of the differing rubber manufacturing companies around the world. It has already been pointed out that the rubber industry is heavily dependent on the automotive industry. It is reasonable to suppose that rubber manufacturers operating in countries with a healthy automotive industry should find it that much easier than companies working where the automotive industry is declining or non-existent. Even within a single trading community, one might also expect wide variations in company success. Companies that fail to do the necessary development work to ensure that their products find a ready market and companies that make inefficient use of manpower, equipment and materials would not be expected to enjoy a good long-term prosperity.

In the main, the rubber industry makes parts that are sold to other

18   J. A. BRYDSON

TABLE 5
ESTIMATED CHANGES IN THE IMPORTANCE OF END-
USE INDUSTRIES IN WESTERN EUROPE

	% 1978	1985
Tyres	51	45
Other automotive products	16	15
Construction	8	10
Building	7	8
Footwear	5	6
Textile, furniture	3	5
Electrical applications	3	3
Sports and leisure goods	2	3
All others	5	5

industries, the car industry being the most well-known case. Table 5 shows a breakdown of use of rubber products into various areas of application in western Europe, both in 1978 and as a prediction for 1985.

The data in this table is, of course, for the relative importance of the various end-uses and a drop in the figure between 1978 and 1985 does not necessarily imply a decline in total tonnage. Some ideas as to whether or not this does imply such a decline may be gauged by considering the predictions of the estimated total rubber consumption in major user countries over the next 10 years (Table 6).[8]

TABLE 6
ESTIMATED TOTAL RUBBER CONSUMPTION IN MAJOR WESTERN EUROPEAN COUNTRIES FROM 1979 TO 1989 (THOUSANDS OF TONNES)

	1979	1984	1989
West Germany	593	678	755
Italy	359	468	552
France	432	480	513
UK	431	480	510
Spain	218	238	276
Belgium/Luxembourg	104	118	137
The Netherlands	67	82	92
Sweden	51	75	85
Austria	53	46	78
Other western European countries	183	199	212

10. CONCLUSIONS

For some years now the rubber industry has suffered from the fact that there have been few obvious new markets to be tapped. Further growth in existing markets is highly dependent on the global economic situation and, to some extent, on increasing competition from plastics materials. Worldwide, the rubber industry has to date coped remarkably well with the catastrophic rises in petroleum prices in the 1970s, although some countries have suffered more than others. The prognosis is that the rubber industry will have to work harder and more efficiently in the next 10 years than heretofore. Those companies that are best equipped to do so will do well but past success will be no guarantee of future prosperity.

REFERENCES

1. Data extracted from *Rubber Statistical Bulletin*, January 1980.
2. Data based on Table 51 from *A Profile of the European Rubber Industry and its Likely Future*. Information Research Ltd, London, 1980.
3. WILKINS, G. Rubber in the motor car, *Eur. Rubber J.*, **161**(10), 1979, 8.
4. ALLEN, P. W. *Plast. Rubber Int.*, **4**, 1979, 161.
5. INTERNATIONAL INSTITUTE OF SYNTHETIC RUBBER PRODUCERS. Forecasts issued Autumn 1979 (global and CPEC data); February 1980 (USA data).
6. SWEET, G. C. Special purpose elastomers in *Developments in Rubber Technology —1*, A. Whelan and K. S. Lee (Eds), Applied Science Publishers, London, 1979, Chapter 2.
7. Editorial statement. *Eur. Plast. News*, **7**(1), 1980, 30.
8. Editorial statement. *Elastomerics*, July 1978, 17.

Chapter 2

STYRENE–BUTADIENE RUBBER

J. A. Brydson

National College of Rubber Technology,
The Polytechnic of North London, Holloway, London, UK

SUMMARY

The historical development of SBR including trends in manufacturing methods is outlined. Variables in micro- and macro-structure are analysed and the properties of the polymer are compared with those of natural rubber. The processability of SBR is considered in some detail and attention is given to recent methods of assessing mixability and other processing characteristics.

Compared with natural rubber SBR compounds are lacking in green strength. Recent developments using labile crosslinking systems to reduce this deficiency are described. The general vulcanisation behaviour of SBR is reviewed and the network structures of the vulcanisates are compared with those of natural rubber. Whilst the chapter is primarily concerned with emulsion SBRs, which constitute the bulk of the market, the relative merits and demerits of the newer solution polymers are considered.

1. INTRODUCTION

Styrene–butadiene rubber was first prepared about 1929 in attempts to overcome the deficiencies of free-radical emulsion polymerised polybutadiene. At the time the advantages gained were not sufficient to lead to commercial use and little progress was made. In 1937, however, in correct anticipation of future difficulties in obtaining the natural material, commercial production was initiated in Germany with the product being

21

marketed as Buna S. More interesting, in 1930 there occurred an event which was to have considerable repercussions on both the rubber industry and on the world at large. In 1930 the German chemical combine I.G. Farbenindustrie had made an agreement with the Standard Oil Co. of New Jersey to assist each other in developing chemical products and processes from petroleum. As a consequence, I.G. Farbenindustrie supplied Standard Oil with the technical details of Buna S production. When, in 1942, the Japanese entered the Second World War and cut off natural rubber supplies to Britain and the United States, the latter country initiated a crash building programme. This led to the construction between 1942 and 1944 of 87 factories with a total annual production of about 1 000 000 t of styrene–butadiene rubber, then known as GR–S (Government Rubber Styrene). The role that this programme played during the war is difficult to over-rate and without it the course of the war would surely have been very different.

At the end of the war, natural rubber again became available and demand for the generally inferior GR–S slumped considerably. Nevertheless, research continued and in the early 1950s styrene–butadiene rubbers became available that had been polymerised at temperatures as low as 5 °C. These *cold rubbers* as they were called, had markedly better properties than the earlier styrene–butadiene (*hot*) rubbers which had been polymerised at somewhat higher temperatures and were in a number of respects, superior to the natural product. With the advent of the Korean War, the importance of major industrial nations becoming independent of the products of a narrow geographic area were soon realised and styrene–butadiene rubbers became re-established in importance. At about this time, the designation GR–S disappeared, to be replaced, eventually globally, by the designation SBR. In due course, SBR was to overtake natural rubber as the major general purpose rubber, at least in tonnage terms, throughout the world.

In the early 1960s, a novel class of styrene–butadiene rubber made its appearance. This type of rubber was prepared using anionic catalysts such as butyl lithium. These rubbers were prepared in solution and hence became known as solution polymers in contrast to the earlier emulsion polymers. Today, the bulk of commercial materials are still of the emulsion type but the potential of the solution rubbers is being increasingly recognised and these may be expected to increase in importance as the years go by.†

---

† The SBS thermoplastic elastomers can be considered as solutions SBRs but consideration of these materials, with their distinct technology, is not included in this chapter.

2. PREPARATION OF STYRENE–BUTADIENE RUBBER (SBR)

Nowadays, both monomers are made from petroleum. However, it is interesting to note that during the Second World War, both Russia and the United States produced butadiene from grain alcohol. At about the same time, German chemists developed a number of synthetic routes from coal via acetylene. Yet another route to butadiene started with oat husks, from which were obtained pentosans and in turn, furfuraldehyde, tetrahydrofuran and butadiene. Styrene may also be produced from a number of materials, particularly from coal.

2.1. Emulsion Polymerisation

Until the early 1950s, the major method of emulsion polymerisation involved water-soluble initiators, such as potassium persulphate, being used to initiate polymerisation in an emulsion system stabilised by a fatty acid soap. Molecular weight was controlled by the use of a mercaptan and polymerisation proceeded at about 50 °C until approximately 72 % of the monomer had been converted into polymer. This process yielded the so-called hot rubbers. Today, the bulk of SBR materials are prepared using so-called redox initiators which comprise a reductant such as ferrous sulphate with sodium formaldehyde sulphoxylate in combination with an oxidant such as p-menthane hydroperoxide.[1] In this case, the polymerisation temperatures are as low as 5 °C and conversion of monomer to polymer is only about 60 %. Both the hot and cold rubbers are taken to number average molecular masses (molecular weights) of about 100 000, unless they are being used for oil extension (see later).

There have been few changes in the basic chemistry of emulsion polymerisation processes in recent years, the main changes being in the finishing stages and in computerisation of the process.[2] Amongst the changes in the finishing processes have been alterations in the method of drying. Whilst it was common to dry the polymerised rubber by a combination of vacuum and hot air oven, it is now more common to use screw dewaterers which effectively squeeze the water out. This has resulted in more efficient and quicker drying. This is followed by the use of extruder driers where the partially dried squeezed rubber crumb is extruded through a die plate. In this process, the moisture flashes off as soon as the rubber is extruded. One other change that has occurred in recent years has been the discontinuation in the use of phenyl-β-naphthylamine as a stabiliser because of the carcinogenic hazards of some of the impurities which may be present.

2.2. Solution Polymerised SBRs

These polymers are produced using anionic catalysts, usually butyl lithium. Besides being ionic, rather than free-radical in character, these polymerisations differ from conventional emulsion polymerisations in a number of important respects. For example, there is no need to wait for the initiating species to decompose into active components as happens with peroxides. For this reason, all of the catalyst molecules can be made active from the beginning of the reaction. In addition, each polymer molecule will continue to grow as long as there is monomer present. In fact, if reactions are carried out under scrupulously clean conditions, it is possible to reactivate polymer growth some time after the initial polymerisation reaction by adding further amounts of monomer. For this reason, the term *living polymer* has sometimes been used to describe these materials. Clearly, it is not necessary for the added monomer to be of the same species as the initial material and this provides a convenient way for making block copolymers. The fact that chains start to grow at about the same time and continue to grow until all the monomer is consumed, results in a narrow molecular mass distribution for solution polymers.

Whereas with emulsion polymerisation using free-radical initiators, the relative reactivities of styrene and butadiene to the growing polymer radicals are not greatly different, thus leading to a reasonably random polymer structure, it is not necessarily the case with solution polymerised materials. In this case, the relative reactivities of the styrene and butadiene to the growing polymer anion, depend strongly on the solvating power of the medium. For example, the addition of styrene is very slow in a hydrocarbon medium but the addition of butadiene is high. Thus, styrene is virtually excluded from the initial polymer formed but becomes increasingly incorporated as the butadiene becomes depleted. Hence, one end of the chain is rich in butadiene and the other end rich in styrene. This is referred to as a *tapered two-block binary copolymer*. However, addition of small quantities of ethers or amines to the polymerisation system markedly changes the situation and increases the reactivity of the styrene. At the same time, such materials do tend to promote a higher vinyl content in the butadiene component which is generally undesirable. An alternative approach to producing a more random chain structure is to use metal butoxides in conjunction with butyl lithium catalyst. A further alternative approach is to employ incremental addition of the butadiene monomer.

The fact that the polymers have terminal anions, enables these molecules to be linked together at their ends. If the linking agent is difunctional, such as in the case of an alkyl dihalide, it is referred to as a coupling agent and the process is known as coupling.[3] Other linking agents have a functionality of greater than two, for example, phosphorus triiodide is trifunctional, silicon tetrachloride is tetrafunctional and divinyl benzene can have functionalities ranging from 7 to 14. In these cases, polymers can be produced which are T-shaped, X-shaped, and even star shaped. Tin coupling agents have also been used which are sufficiently weak to be broken down during mastication operations. It will thus be seen that there is a substantial flexibility in the solution polymerisation process which will be discussed subsequently.

3. STRUCTURE OF SBRs

Styrene–butadiene rubbers vary in both macro-structure and micro-structure[4] and these variables can have important influences on the properties of the polymer.

Macro-structural variables include:

1. Monomer ratio, including the possibility of additional monomers
2. Average molecular mass
3. Molecular mass distribution
4. Linearity or otherwise of the molecular structure
5. Extent of gel

Micro-structural variables include:

1. *Cis-/trans-*/vinyl group ratios in the butadiene segment
2. Distribution of styrene and butadiene units

3.1. Macro-structural Variables

Butadiene and styrene may be copolymerised in any ratio and, in many respects, physical properties change in a linear or near-linear manner with the fraction of styrene (or butadiene) residues. Thus while both *cis-* and *trans-*1,4-polybutadienes have T_gs of about $-100\,°C$ and polystyrene a T_g of about $+90\,°C$, a 50:50 copolymer has a T_g of about $-5\,°C$. (This is, however, influenced by varying the extent of pendant vinyl groups through 1,2-polymerisation of the butadiene units, since these act rather like the pendant benzene rings in causing stiffening of the main chain.) Most emulsion polymers have styrene contents of about 23.5% and solution polymers about 25%, these being generally believed to give the best properties. Additional monomers are hardly ever used, although divinyl benzene is used in some 'hot' rubber grades to produce a material which, when blended with conventional rubbers has the effect of reducing die swell (types 1009 and 1018).

The non-oil-extended grades usually have $\bar{M}_n$ values of about 100 000. Lower values lead to polymers that show excessive bale distortion on storage, whilst high values lead to increased difficulties in processing. There is a paucity of published information on the effect of the molecular mass of the base polymer on vulcanisate properties. However, one treadwear study using a polymer with a range of Mooney values showed that molecular masses (in the range under consideration) did not have an important effect.[33]

In commercial practice, it is common to polymerise to a higher level and then incorporate an oil, typically $37\frac{1}{2}$ phr to reduce the compound viscosity. This will yield a compound with similar properties to one based on a polymer of lower molecular mass (without oil) but at a lower cost. Such oil-extended SBRs are of great importance to the tyre industry.

Molecular mass distributions are much higher with emulsion polymers than with typical solution polymers. This is most simply expressed by the $\bar{M}_w/\bar{M}_n$ ratio which is in the range of 4–6 for 'cold' emulsion grades and 1·5–2 for typical solution rubbers. This can have an important effect on processability as will be discussed later.

Branching is usually greater with emulsion polymers than is really desirable. Solution polymers may be either completely linear or contain a controlled degree of branching. The two polymer types are also different in the amount of gel that is present. Emulsion polymers contain micro-gel (extensively in hot rubbers and moderately in cold rubbers) as a result of the very nature of the free-radical emulsion polymerisation process, whereas the gel content of solution polymers is very small. Rather interestingly, there have been deliberate attempts recently to incorporate controlled amounts of gel into the SBR in order to improve polymer green strength as will also be discussed later (see Section 5.2).

3.2. Micro-structural Variables

Cis-/trans- ratios (usually less than 1) do not seem to be a very important variable in determining the mechanical properties of vulcanisates. On the other hand, the vinyl content (typically about 12 % in an emulsion rubber) may be expected to be important. For example, there is a near linear relationship between vinyl content and T_g with polybutadiene and this effect may be expected to also occur with SBRs.

More important is the distribution of styrene and butadiene units. It is possible to discriminate five basic types of styrene–butadiene distribution.[5]

These are:

1. Uniform random copolymers. In this case the monomer distribution closely follows that expected from random statistical considerations and does not vary either along the length of the polymer chain or between chains. Emulsion polymers generally approximate to this class.

2. Random copolymers that show inter-chain variations but uniform composition within a chain. This situation is possible with emulsion polymerisation but not with solution polymerisation (unless resulting from blends of different batches). It occurs where reactivity ratios are different and where no allowance is made to control the monomer feed ratio.

3. Random copolymers which show variations between chains and along the chain. This is likely to occur with solution polymerisation where the reactivity ratios differ widely and where no monomer feed control is exercised.

4. Block copolymers consisting of two or more block sequences, at least one of which is not a homopolymer or of a uniformly random copolymer sequence. In solution polymerisation, these can occur where a second monomer is added before the first has been completely consumed.

5. Ideal block copolymers in which all block sequences are uniform in composition (either homopolymeric or uniformly random copolymeric) and which change discontinuously at the junction points. Such systems are produced by solution polymerisation where a new monomer is not added until the first has been totally consumed. They may also be obtained, at least in principle, by coupling two differing polymer species. In practice, coupling of like species is also likely to occur simultaneously.

4. GENERAL PROPERTIES OF SBR

Before discussing them in detail, it is instructive to summarise the properties of SBR into two categories:

(a) Those properties in which they are similar to natural rubber
(b) Those properties in which they are distinct from natural rubber

4.1. Similarities

Like natural rubber, SBR is an unsaturated hydrocarbon polymer. Hence, unvulcanised compounds will dissolve in most hydrocarbon solvents and other liquids of similar solubility parameters,[4] whilst cured stocks will swell extensively. Both materials will also undergo many olefinic type reactions such as oxidation, ozone attack, halogenation, hydrohalogenation and so on, although the activity and detailed reactions differ because of the presence of the adjacent methyl group in the natural rubber molecule. Both rubbers may be reinforced by carbon black and neither can be classed as a heat-resistant rubber.

4.2. Differences

The differences between the rubbers can be subdivided into three categories:

(i)   In the materials supplied
(ii)   In processing behaviour
(iii) In the properties of the vulcanisate

Compared with the natural material, raw SBR is more uniform in many senses. It is more uniform in quality and compounds are more consistent in both processing and product properties. It is more uniform in the sense that it usually contains fewer undesired contaminants. Finally, over the years it has been found to be more uniform in price with smaller market fluctuations. It must, however, be said that the advent of improved grades of natural rubber such as the SMR types has reduced the difference in recent years.

A major difference between SBR and natural rubber is that the former does not break down to any great extent on mastication. The synthetic material is supplied at a viscosity considered to provide the best balance to give good dispersion of ingredients and ease of flow in extrusion, calendering and moulding. This provides savings in both energy consumption and time, and hence on costs. In addition, it is easier to work in reworked stock which is changed little in properties from the original material. On the debit side, it is to be noted that mill mixing is usually less easy than with natural rubber and that sulphur vulcanisation is somewhat slower and requires more powerful accelerator systems. More important still, the synthetic rubber lacks tack and green strength and this is of consequence in tyre building, particularly with radial tyres.

Whilst natural rubber is crystalline with a T_m of about 50 °C, SBR is amorphous. Although crystallinity in natural rubber is reduced by the presence of crosslinks and of fillers and other additives, it still crystallises on extension giving a rubber of good tensile strength, even with gum stocks. On the other hand gum vulcanisates of the amorphous SBR are weak and it is essential to use reinforcing fillers such as fine carbon blacks to obtain products of high strength. Black-reinforced SBR vulcanisates do, however, exhibit very good abrasion resistance, generally being superior to comparable natural rubber compounds above 14 °C. On the other hand, they have lower resilience and resistance to tearing and cut growth. The ageing behaviour of SBR is quite different to that of natural rubber with the synthetic material tending to crosslink rather than exhibit chain scission on oxidation.

5. PROCESSABILITY

Successful processing of SBR requires attention to several aspects. These include:

1. The ability to incorporate fillers such as carbon black easily
2. Good mill banding characteristics and resistance to batch break-up in internal mixers
3. Good flow characteristics including ability to extrude from dies with sharp edges
4. Control of die swell
5. Optimisation of green strength
6. Provision of a suitable level of tack

5.1. Testing for Processability
The above aspects are very different characteristics related in many cases to different molecular features and it is unreasonable to expect that one single test can give an overall measure of processability. It is therefore surprising that efforts to devise such a simple test have continued to be made and perhaps, even more surprising, that some measure of success has been achieved.

5.1.1. Use of the Brabender Plastograph
'Mixability' is generally concerned with the readiness of the rubber to take up substantial quantities of carbon black or other fine filler, usually in an internal mixer. One quantitative way of studying this is to follow the power

consumption pattern during internal mixing. Beach *et al.*[6] suggested that a useful criterion was the time interval between loading the mixer and observing the second peak in the power consumption–mixing time curve. It was found that rubber–carbon black blends mixed for this time had similar levels of black dispersion in the rubber and the term 'black incorporation time' was suggested. The use of an internal mixer for such studies is clearly not an ideal laboratory situation and subsequently Meder and May[7] found that similar data was obtainable using a Brabender Plastograph, in this case measuring the time to the second peak in the torque–time curve. This test is now commonly referred to as the BIT test.

5.1.2. Mooney Viscometer

Meder and May also studied the use of the Mooney viscometer, more common in rubber laboratories than Brabenders, for assessing mixability. They reviewed various methods which involved recording changes in Mooney viscosity during mixing. Some workers[8] recommended noting the change in Mooney value between 1 and 15 min,[8] others the difference between the maximum and minimum Mooney value. In general, the differences are known as the Delta Mooney (ΔM) value.[9] Somewhat surprisingly, it was found that ΔM values depended on storage conditions while they were also subject to larger inter-laboratory variations than the Brabender test. Einhorn[10] suggested that the Mooney viscosity (ML 1 + 4 at 100 °C) of a compound subjected to a short Banbury cycle gave a measure of processability which he called the 'Banbury processability index'.

5.1.3. Capillary Rheometer

An attempt was made by Tokita and Pliskin[11] to discover why the second peak in the torque–time curve of the BIT test should correspond to a particular level of black dispersion. In the course of this work it was found that die swell also varied with extent of dispersion and peaked at a similar dispersion level. This provided an alternative method to assessing BIT for laboratories possessing capillary rheometers but not Brabenders. The test, however, is more time consuming as it does require several mixes to be made at various levels of dispersion. In their paper, Tokita and Pliskin related processability, including mixability, to various viscoelastic parameters including the extensibility of the raw rubber. In turn, these were related to molecular parameters. For example, they showed that too narrow a molecular mass distribution leads to a low value of extensibility and that this can cause break-up during mixing. As a result of their studies they concluded that, compared with a general purpose emulsion SBR, as

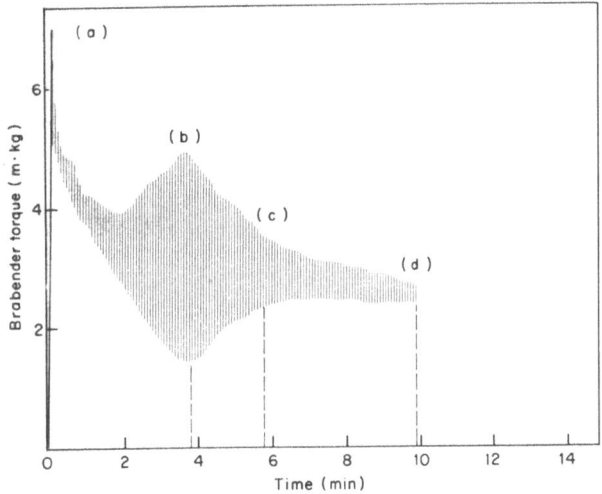

FIG. 1. Brabender Plastograph mixing curve for 137·5 parts SBR 1712 plus 69 parts N-339, mixed at 50 rpm, 138 °C, 'upside down' (black added first). Curve area (area below top trace) = 3·9 m kg min, $W_u = 2120\,MJ/m^3$. (a) Ram lowered, (b) black incorporation time or second power peak, (c) t' point, (d) dump. Reproduced with permission from *Rubber Chem. Technol.*[13]

currently available, it would be better, in terms of processability, to use polymers with the compromise of:

1. A slightly narrower molecular mass distribution
2. A more linear structure
3. A lower styrene content
4. A lower Mooney viscosity

5.1.4. Mixing Work Input

Van Buskirk *et al.*[12] claimed that the flow behaviour of SBR–black compounds was a function of mixing work input. Flow behaviour was independent of mixer size, speed and mixing time as long as the temperature–time profiles were identical. From this they introduced the unit work concept. In a later paper Turetzky *et al.*[13] suggested that rather than using the second peak of the torque–time curve as in the BIT test described above, it would be more appropriate to take a later point on the torque–time curve, the so-called t' point. The position of this t' point is illustrated in Fig. 1.

These workers noticed that increasing the molecular mass of the rubber

TABLE 1

CORRELATION COEFFICIENTS BETWEEN MOLECULAR WEIGHT PARAMETERS AND PROCESSABILITY TESTS (SOLUTION POLYMERISED OIL-EXTENDED SBR)[16]

| | | | | | Correlation Matrix | | | | |
	$\bar{M}_n$	$\bar{M}_w$	$\bar{M}_z$	$\bar{M}_n/\bar{M}_n$	BIT	BIT unit work	t' Point	t' Point unit work	Die swell
ML$_4$	0.19	0.47	0.27	0.01	-0.10	0.01	0.12	0.27	-0.12
PMT	0.61	0.37	-0.09	-0.52	-0.66	-0.60	-0.40	-0.30	-0.69
Δ Mooney	-0.78	-0.30	0.36	0.75	0.92	0.85	0.74	0.67	0.87
Mooney slope	-0.78	-0.09	0.59	0.83	0.87	0.87	0.68	0.65	0.89
Die swell	-0.86	-0.23	0.46	0.86	0.87	0.85	0.74	0.79	
BIT	-0.81	-0.28	0.38	0.80					
BIT unit work	-0.81	-0.15	0.47	0.85					
t' Point	-0.76	-0.32	-0.23	0.73					
t' Point unit work	-0.72	-0.25	0.29	0.73					

PMT = Initial peak Mooney torque time.

increased the time to reach the power peak and also the corresponding die swell peak. The quality of black dispersion was found by Mills *et al.*[14] to be very sensitive to molecular mass distribution but they found that neither the die swell peak nor the Mooney viscosity was much affected. Smith[15] found that there was no simple relationship between Mooney viscosity and processability.

5.1.5. Comparison of Tests

By 1976, a variety of tests had therefore become available, many developed in Australia, and these were subjected to what might be considered a definitive comparison by Mills and Giurco,[16] also working in Australia. They compared Delta Mooney, BIT tests, die swell peak times and t' points. Table 1, taken from Mills and Giurco's paper, gives the correlation coefficients, not only between some of the processability tests but also with a number of molecular parameters. (A value of 1 indicates a perfect positive correlation, a value of 0, zero correlation and a value of -1, perfect negative correlation.)

Of particular interest in the results is that once again a good correlation between BIT and the die swell peak time is observed and also that both of these times also increase with increase in molecular mass distribution. It would therefore seem that, whilst the BIT may give a time rather less than required for complete mixing, it does consistently appear to provide a means of rapidly rating rubber in order of mixability.

A rubber technologist will also have some interest in obtaining a measure of the flow properties of his compounds. Traditionally, this has been achieved using the ubiquitous Mooney viscometer. This has the virtue that a single point reading may be quickly and easily obtained. It suffers the disadvantage that it gives an assessment viscosity at shear rate probably much less than used in many operations such as injection moulding so that there has recently been increased use in capillary rheometry for measuring shear stress–shear rate relationships. Flow curves obtained from such equipment should, however, be assessed with care[17] because of possibilities of wall slip, shear heating, and other factors. Very recently, Palit and Bar-el[18] have found that if the SBR–black compounds are heated (in the absence of a curing system) at different temperatures before actually carrying out the flow tests, different flow curves are obtained.

Die swell measurements can also be made using the capillary rheometer and Fig. 2, from the Mills and Giurco paper,[16] shows the dependence of die swell on the molecular mass distribution for emulsion polymerised SBRs and solution polymerised SBRs.

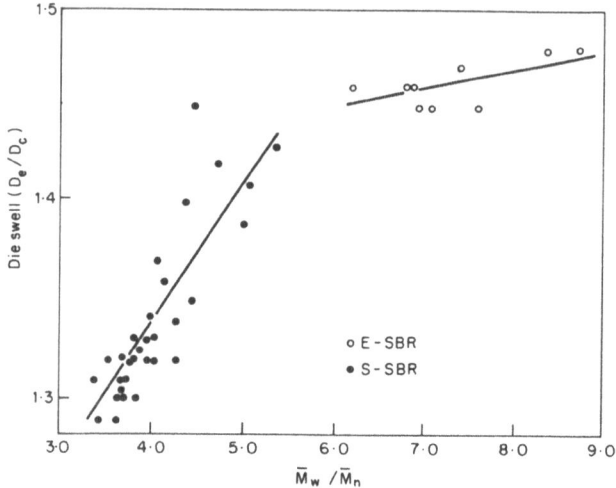

FIG. 2.   Influence of polydispersity on die-swell of oil-extended SBR compounds.
D_e = extrudate diameter, D_c = capillary diameter. Reproduced with permission
from *Rubber Chem. Technol.*[16]

An alternative approach to the use of a conventional rheometer is to use a
Brabender Plasti-corder or Plastograph with an extruder rather than a
mixing attachment. These could be used to obtain flow curves whilst
alternatively, a Garvey die may be fitted onto the extruder head and
extrudability (as assessed by the ability to extrude sharp edged products
without feathering) may be studied.

5.2. Polymer Modification to Enhance Green Strength

The development and general acceptance of the radial tyre has led to
attention being paid to a property known as *green strength*. This is a quality
of particular importance in carcase stocks for radial tyres since these may
be subjected to deformations up to three times the original dimensions
during the forming stage.

5.2.1. Characterisation of Green Strength

A high green strength does not simply imply high tensile strength or
stiffness but is usually taken to imply a quality of cohesiveness in stretched,
uncured rubber compounds which prevents thinning down and breaking
short, cord strike through and instability during fabrication.[19] Whilst there
is no single test that completely characterises green strength, the shape of

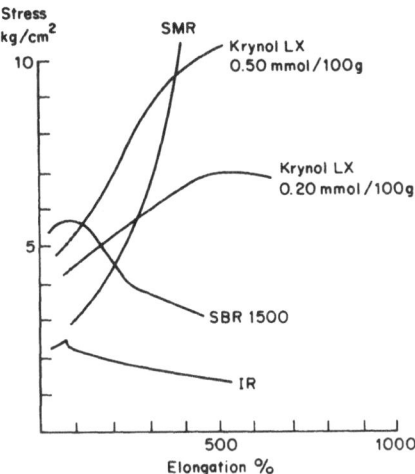

FIG. 3.   Instron stress–strain curve on cohesive rubber compounds with 50 phr N550 black (40 °C, 50 cm/min). Reproduced with permission from *Elastomerics*.

the tensile stress–strain curve provides a useful indication. Thus, whereas the curve for uncured natural rubber compounds is continuously positive, that for SBR compounds becomes negative at elongations of about 100 % (Fig. 3). Some workers[20] specifically suggest that change in stress in elongating the sample from 200 to 300 % can give a quantitative assessment of green strength.

5.2.2. Labile Crosslinking

The difference between natural rubber and SBR compounds is generally accepted as being due to the crystallisation that occurs in natural rubber but not SBR on extension. Buckler *et al.*[19] have pointed out that this crystallisation in natural rubber could make the rubber quite unprocessable were it not for the fact that polyisoprene chains readily break down under shear. Whilst it was recognised that SBR could not be made crystalline, the above considerations led Buckler and his colleagues at Polysar to develop SBR-type polymers that were lightly crosslinked but with the crosslinks being weaker than the backbone bonds of the polymer chains. Thus, during mixing and extrusion, these crosslinks would break preferentially and give a material similar to regular SBR compounds. Furthermore, such labile crosslinked systems were of a type that could be processed at room temperature so that after such operations as mixing and extrusion, the crosslinks would again be formed during a so-called resting period.

This approach has continued to be developed by chemists[20,21] at Polysar and they have paid particular attention to the use of terpolymers, known as SBR-LX rubbers, based on styrene, butadiene and a third monomer containing *t*-amine groups. Crosslinking via the *t*-amine groups may be brought about by adding a dihalide at the latex stage in order to give crosslinks involving quaternary salts (Fig. 4). A typical polymer of this type

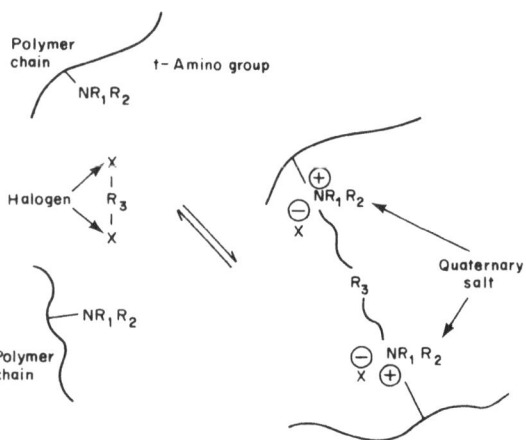

FIG. 4.   The labile crosslinking reaction for enhancing green strength. Reproduced with permission from *Rubber Chem. Technol.*

possessed about 18 amino groups per weight average molecular mass chain, which typically had a weight average molecular mass of 400 000. The amount of crosslinking agent was such as to give about two crosslinks per weight average molecular mass chain. The stress–strain curves of unvulcanised oil-extended compounds of these materials (designated Krynol LX) were found to be much closer to those of corresponding natural rubber compounds than of compounds based on conventional SBR (Fig. 3).

These crosslinks can be broken by mechanical shear but re-form in a matter of days at room temperature or minutes at 160 °C. It was, however, observed that intensive shear did lead to a certain amount of irreversible viscosity drop. This has been ascribed to mechanical rupture of very high molecular mass chains which before mechanical shear have a disproportionate influence on the network structure.

Such irreversible viscosity changes dependent on shear, as well as other factors, make it difficult to simulate factory conditions using laboratory

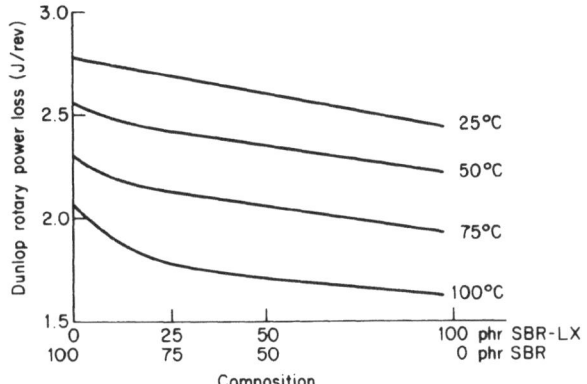

FIG. 5.   SBR-LX experimental trends. Power loss versus SBR-LX content.
Reproduced with permission from *Elastomerics*.

tests and the importance of factory trials is stressed by those developing the
process. More recent work in this area has also emphasised the usefulness of
blends of conventional SBR and polybutadiene with the new terpolymers.
One particularly important observation is that tyre rolling resistance, as
assessed by a Dunlop Rotary Power Loss Machine appears to go down with
increasing amounts of crosslinked polymer,[20] a most desirable feature
because of the need for improved fuel economy (Fig. 5).

6.   VULCANISATION OF SBR

6.1. Rate of Cure
SBR is slower curing than natural rubber and higher accelerator levels are
necessary in order to obtain equivalent cure times. This simple statement
does, however, mask the fact that the cure rates of different grades of SBR
differ considerably. This is illustrated in Fig. 6 for a series of Cariflex grades
produced by Shell,[22] although this data does not take into account the
variation in rubber hydrocarbon content between grades. It is seen from
Fig. 6 that a S-1500 type of SBR requires a 50 % higher accelerator loading
than a S-1712 rubber to achieve an optimum cure of 30 min at 144 °C using a
standard test recipe (Table 2). Furthermore, where compounds are adjusted
to have similar optimum cure time, the relative scorch times may be quite
different, as illustrated in Fig. 7.

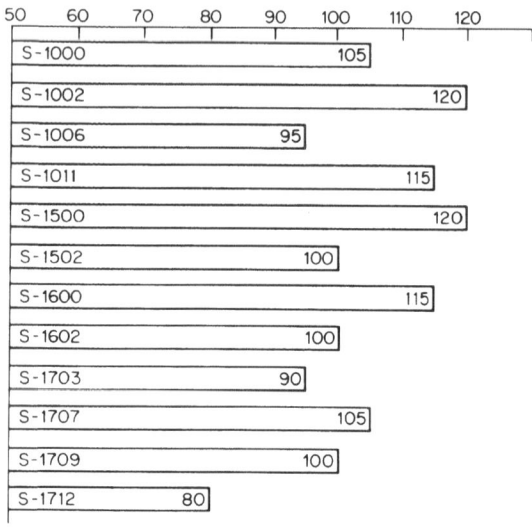

FIG. 6.   Relative accelerator levels for various SBR types to give optimum cure in the same cure cycle (level for S-1502 represented by 100). Data is for Shell Cariflex grades which follow standard ASTM nomenclature:

1000 series: 'Hot' emulsion grades
1500 series: 'Cold' emulsion grades
1600 series: Black masterbatch of 'cold' grades
1700 series: Oil-extended 'cold' grades
1800 series: Oil-black masterbatch of 'cold' grades

Oil-extended polymers considered 100% rubber hydrocarbon. Reproduced by courtesy of Shell Chemical Co.

6.2. Effects on Vulcanisate Properties
It is perhaps also worth stressing that different properties change with cure time in different ways. Such differences are illustrated schematically in Fig. 8. Figures 9, 10 and 11, show specifically the effect of altering the type and amount of acceleration on the optimum cure time, tensile strength and modulus of compounds based on oil-extended SBR.

6.3. Crosslink Types and Network Structure
In the 1960s, considerable work was undertaken by the Malaysian Rubber Producers Research Association and its forerunners to study the nature of sulphur crosslinks in natural rubber vulcanisates. It was found that whereas conventional accelerator–sulphur systems led to substantial amounts

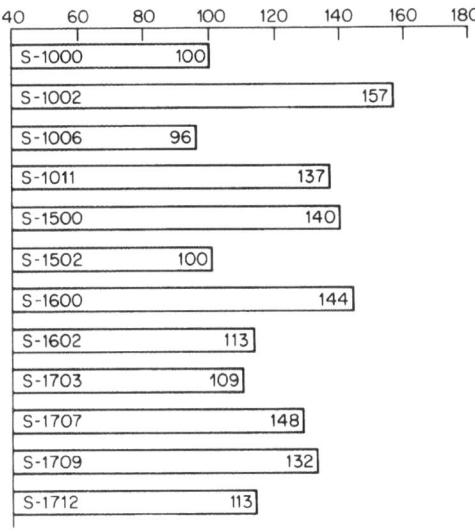

FIG. 7. Relative scorch times for SBR types (time for S-1502 represented by 100).
Reproduced by courtesy of Shell Chemical Co.

TABLE 2
TEST RECIPE
(Reproduced by courtesy of Shell Chemical Co.)

	phr	phr
Polymer: Cariflex S-1000, S-1500, S-1700 series	100·0	—
Black masterbatch; Cariflex S-1600 and S-1602	—	150·0[a]
Zinc oxide	3·0	3·0
Stearic acid	2·0	2·0
PBNA	1·0	1·0
HAF black	50·0	—
Dutrex 20	5·0	5·0
Sulphur	2·0	2·0
Santocure 1·0 ⎱		
TMTD 0·1 ⎰	Variable[b]	

[a] Cariflex S-1600 and S-1602 contain 50 phr HAF black.
[b] The accelerator level was adjusted to produce an optimum cure at 30 minutes at 292 °F.

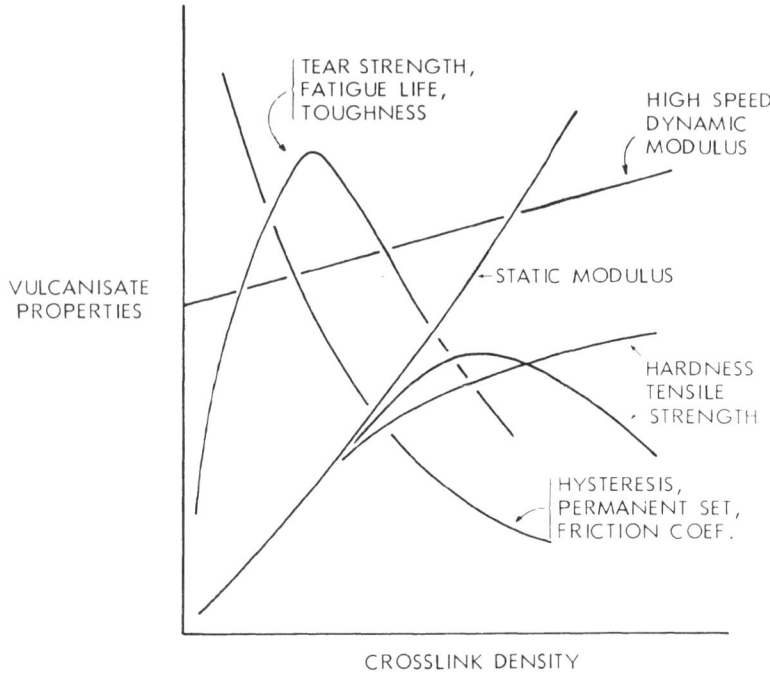

FIG. 8.   Dependence of vulcanisate physical properties on crosslink density.[23]

of disulphide and polysulphide linkages, there were very few monosulphide links. On the other hand, if the sulphur content was reduced, and the accelerator content boosted, the vulcanisates had a much higher proportion of monosulphide crosslinks. In the case of natural rubber, the higher monosulphide content led to a distinct improvement in resistance to thermal and oxidative ageing but at the expense of a much reduced flex life. Such high accelerator–low sulphur systems are frequently referred to as efficient vulcanisation systems (EV systems), the adjective in fact only implying a more efficient use of sulphur than necessarily that of a better product.

Subsequent studies using SBR have shown several interesting features which have been dealt with at length in the first monograph[23] of this series but which may be briefly summarised as follows:

1.   Conventional SBR vulcanisates have a monosulphide crosslink content similar to that of an EV natural rubber system (Table 3).

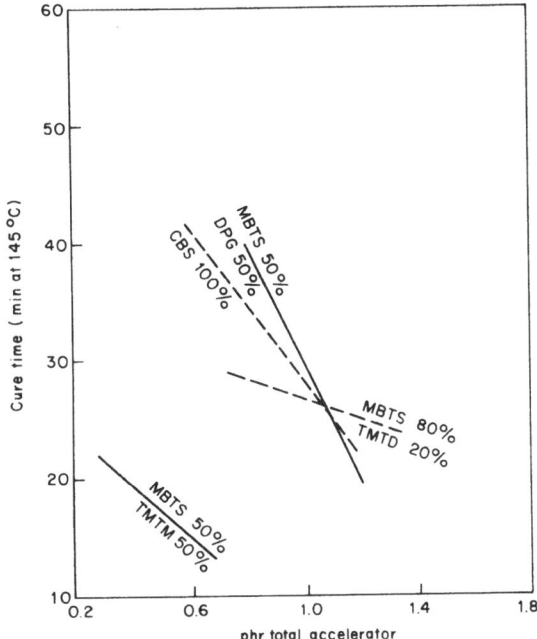

FIG. 9. Dependence of optimum cure time on accelerator type and concentration using a compound based on oil-extended 'cold' rubber (type 1712) (data from Shell Chemical Co.). Formulation: Cariflex S-1712, 100; ZnO, 3; stearic acid, 2; phenyl-β-naphthylamine, 1; HAF black, 50; mineral oil (Dutrex 20), 5; sulphur, 1·75; accelerators—variable. Reproduced by courtesy of Shell Chemical Co.

TABLE 3
CROSSLINK DISTRIBUTIONS FOR NR AND SBR

| | Crosslink type (%) | | | |
| | NR | | SBR | |
Vulcanisation system	S_1	$S_2 + S_x$	S_1	$S_2 + S_x$
Conventional[a]	0	100	38	62
EV[b]	46	54	86	14

S_1 = monosulphide links; S_2 = disulphide links; S_x = poly-sulphide links.

[a] Conventional for NR: sulphur 2·5, MBS 0·6; conventional for SBR: sulphur 2·0, CBS 1·0.

[b] EV for NR: CBS 1·5, DTDM 1·5, TMTD 1·0; EV for SBR: CBS 1·5, DTDM 2·0. TMTD 0·5.

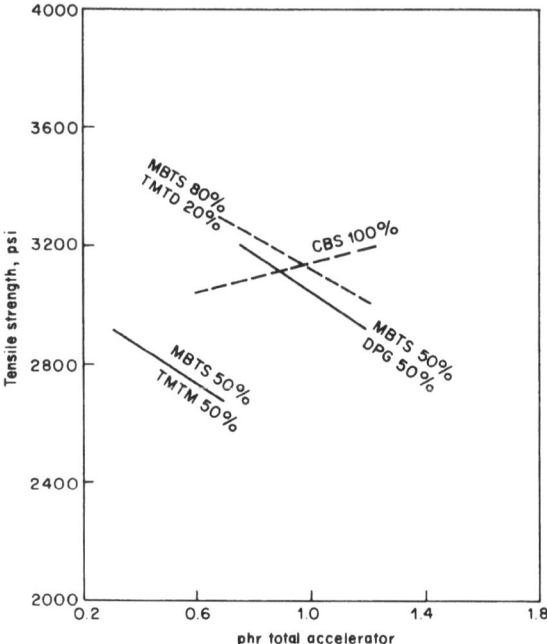

FIG. 10. Dependence of vulcanisate tensile strength on accelerator type and concentration. Formulation as for Fig. 9. Reproduced by courtesy of Shell Chemical Co.

TABLE 4
FATIGUE PERFORMANCE OF SEMI-EV IN SBR

	Conventional	Semi-EV	
CBS	1·2	2·5	1·0
DTDM	—	—	1·0
Sulphur	2·0	1·2	1·2
Fatigue life—80 % extension, kc to failure	323	275	416
Aged 3 days at 85 °C, fatigue life—80 % extension, kc to failure	61	81	117

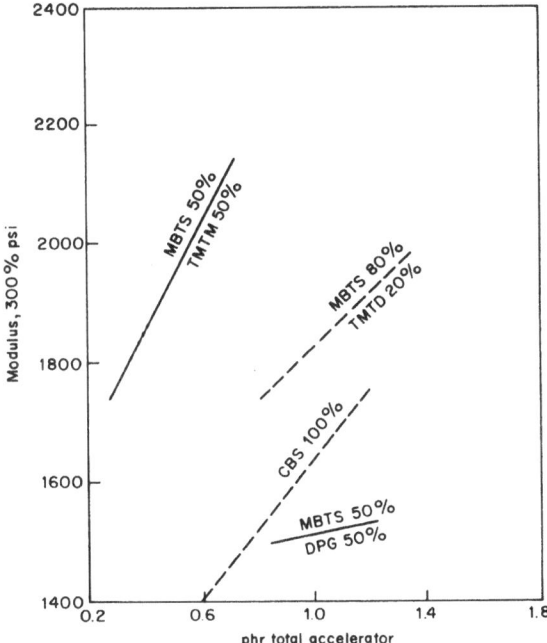

FIG. 11. Dependence of vulcanisate modulus on accelerator type and concentration. Formulation as for Fig. 9. Reproduced by courtesy of Shell Chemical Co.

2. The replacement of a conventional cure by an EV curing system also increases the monosulphide content with SBR, in this case to a value about twice that for a natural rubber EV system (Table 3).

3. Whereas the crosslink density of conventionally cured SBR vulcanisates increases on ageing at elevated temperatures (e.g. 110 °C) the EV-cured material has a very stable crosslink density at the same temperature.

4. The reduction in fatigue life shown with natural rubber EV systems is not duplicated with SBR (Table 4). Indeed, EV SBR compounds after ageing show much better fatigue resistance than conventional compounds after ageing (Table 5).

In general, it would appear that use of EV systems in SBR leads to a lower aged modulus and hardness, better retention of elongation at break and a general overall improvement in compression set and heat build-up. In practice, because of the high cost of accelerators, true EV systems are not

TABLE 5
PROPERTIES OF FULL EV SYSTEMS IN SBR

	Conventional	Full EV	
MBS	1·2	7·0	1·2
TMTD	—	—	1·2
DTDM	—	—	1·2
Sulphur	2·0	0·75	—
Aged 10 days at 90 °C,			
% increase 300 % modulus	115	65	60
Compression set,			
22 h at 70 °C, %	18	11	11
Fatigue–75 % extension			
Unaged, kc	645	750	800
Aged, kc	192	445	400

often used with SBR. Semi-EVs (that is of intermediate accelerator–sulphur ratios) provide a useful compromise between cost and performance.

7.   COMPARISON OF EMULSION AND RANDOM SOLUTION SBRs

7.1. Characteristics of Solution Polymers
From the advent of solution polymerised SBRs it was recognised that they possessed a number of advantages over the emulsion polymers. Subsequent studies[24–26] revealed additional desirable features so that today the following advantages are now recognised:

1. Lighter colour (of many grades)
2. Lower non-rubber content
3. Better dimensional stability of extruded products
4. Faster cure rates
5. Vulcanisates have better resistance to tear, flex cracking and groove cracking
6. Vulcanisates have better abrasion resistance
7. Better low temperature properties

On the other hand the unmodified solution polymers with random monomer distribution generally exhibit:

1. Poor processability (now largely recognised as being associated with a narrow molecular mass distribution)

2. Lower tensile strength
3. Lower modulus

The anionically initiated solution polymerisation methods also have the virtue of considerable flexibility with much greater scope than with the emulsion polymers for ringing the changes in their molecular architecture. This scope was demonstrated by Cooper and Nash[27] who described the various techniques for making polymers of six different types of molecular architecture, namely:

1. Narrow molecular mass distribution—linear
2. Narrow molecular mass distribution—T-branched
3. Broad molecular mass distribution—linear
4. Narrow molecular mass distribution—X-branched
5. Narrow molecular mass distribution—star-branched
6. Broad molecular mass distribution—star-branched

7.2. Improving Properties of Solution Polymers

As a result of this flexibility considerable efforts have been made in recent years to improve those properties in which the solution polymers are inferior to the emulsion materials and also to increase further the superiority of the solution polymers in respect of such properties as abrasion resistance.

7.2.1. Processability

One particular problem was associated with processability. Conventional emulsion and random solution SBRs do not breakdown easily on mastication. Hence typical polymers may be too soft to give adequate shear for good filler dispersion and yet be too stiff for such uses as sponge compounds. Uraneck and Short[28] have described the use of tin coupling agents which join together the living ends of three of four polymer chains to give either a T-shaped or X-shaped molecule. The tin-based coupling site is somewhat weak and may be broken down on shearing in the presence of certain peptising agents such as stearic acid. This concept of peptisable or mechanically degradable materials has been extensively developed by the Phillips Co. and several of their Solprene materials are of this type (e.g. Solprene 1204). An extensive comparison between such polymers and cold emulsion polymers both in oil-extended and unextended forms was made by Railsback and Haws.[26]

7.2.2. Tensile Strength

The marginally lower tensile strength of vulcanisates based on solution

SBRs on occasion causes difficulty in maintaining specifications drawn up with emulsion polymers in mind. In a recent interesting series of experiments Duck et al.[29] prepared a series of blends of solution SBRs of differing average molecular mass but in such proportions as to give polymers with a Mooney viscosity of 50 $\pm$ 10 (ML 1 + 4 at 100 °C). It was found that a number of these blends gave tensile strengths significantly higher than obtained with unblended narrow molecular mass distribution polymers of similar Mooney viscosity.

One such blend consisted of 35 parts of a polymer with $\bar{M}_n$ = 115 000 and $\bar{M}_w$ = 150 000 with 65 parts of a polymer with $\bar{M}_n$ = 257 000 and $\bar{M}_w$ = 656 000. The $\bar{M}_n$ of the blend was 163 000. This blend had a tensile strength of 31·8 MPa compared with a value of 26·1 of similar $\bar{M}_n$ but lower Mooney. For a polymer of similar Mooney (but higher $\bar{M}_n$) the tensile strength was 28·7. These were improvements of 22 and 11 % respectively. This improvement has been related to the presence of some very large molecules of molecular mass in excess of 10^6. It has been suggested that, unlike conventional SBR molecules, these rupture during mixing with the ruptured ends reacting with the surface of carbon black particles to give superior reinforcement.

7.2.3. Other Properties
In an earlier paper Haws et al.[30] studied the effect of changing styrene and

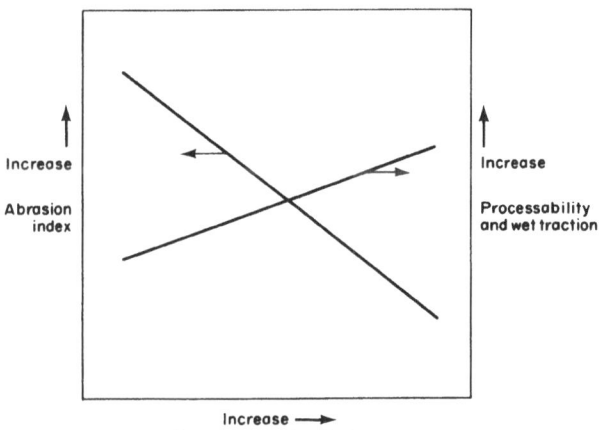

FIG. 12.   General effect of increasing styrene or vinyl concentrations in solution SBRs on abrasion index, processability and wet traction. Reproduced with permission from *Rubber Industry*.[30]

pendant vinyl group content on abrasion index, processability and wet traction. In large measure an increase in abrasion resistance goes hand in hand with a decrease in processability and wet traction (Fig. 12).

A polymer with a butadiene–styrene ratio of 77/23 and a low pendant vinyl group content gave compounds with an abrasion resistance some 10 % superior to corresponding compounds based on an emulsion polymer. The compound also had good processability but inferior tensile strength and tack. Wet traction was almost as good as for the emulsion material.

More recently Livigni et al.[31] have prepared solution SBRs using a barium t-butoxide salt in conjunction with an organo-lithium catalyst. A polymer with a low bound styrene content (9 %) and a high trans- content (70–75 %) was sufficiently regular to undergo strain crystallisation. This phenomenon also occurs with natural rubber and it is interesting to note that the synthetic material exhibited both good tack and green strength.

8.   USAGE OF SBR

Production of SBR now amounts to about 55 % of all synthetic rubber production and 40 % of all rubber production (natural plus synthetic).[32] It is therefore easily the leading type of rubber in terms of tonnage production.

8.1. Major Markets

About two-thirds of this output of SBR goes into tyres but it also finds widespread use in many other areas of rubber goods production such as belting, flooring and mats, hose, shoe soles and heels and many other industrial and domestic applications. In the great majority of these uses the material is used in black-reinforced form.

As pointed out in Chapter 1 the SBR market is facing considerable difficulties. The drop in tyre manufacture and some recovery of natural rubber in this market has led to a situation of over-capacity and low profitability. This is leading to two consequences: (i) Old, relatively inefficient polymerisation plants are being phased out; and (ii) little new plant is coming on stream. In particular this has retarded the development of solution SBRs which are still only about 15 % of the market. Whilst it is widely believed that the solution SBRs are the general purpose rubbers of the future there is little incentive at present to invest heavily in this area until some degree of supply–demand balance has been achieved.

8.2. Competition

More positively there is no substantial evidence that SBR is likely to be replaced extensively by other rubbers in non-tyre applications, this being assured largely by the relatively low cost of the polymer. No new general purpose rubbers appear to be on the horizon at the present time. There may be some replacement where up-grading of product specifications for such properties as ozone and heat resistance requires the use of ethylene–propylene rubbers. On the other hand the advent of the new SBR polymers such as the low styrene strain-crystallising materials described in the previous section with their improved tack and green strength may well, if their initial promise is realised, capture some of the markets currently held by the more expensive natural materials.

REFERENCES

1. URANECK, C. A. *Polymer Chemistry of Synthetic Elastomers: Part I*, J. P. Kennedy and E. G. M. Tornqvist (Eds.), Interscience, New York, 1968, Chapter 4A.
2. GLANVILLE, L. M. and BOWMAN, I. J. Review paper—Progress of general purpose synthetic rubbers. *Progress of Rubber Technology*, Vol. 40, Plastics and Rubber Institute, London, 1977, p. 21.
3. SALTMAN, W. M. (Ed.) *The Stereo Rubbers*, Wiley, New York, 1977.
4. BRYDSON, J. A. *Rubber Chemistry*, Applied Science Publishers, London, 1978.
5. KRAUS, G., CHILDERS, C. W. and GRUVER, J. T. *J. Appl. Polym. Sci.*, 11, 1967, 1581.
6. BEACH, K. C., COMPER, L. F. and LOWERY, V. E. *Rubber Age*, 85, 1959, 253.
7. MEDER, A. and MAY, W. *Rubber J.*, 146 (June), 1964, 39.
8. HEAL, C. J. A. Paper to *Deutsche Kantschuk Gesellschaft, West Berlin*, October 1960.
9. ASTM STANDARD, D-3346-74, July 1975.
10. EINHORN, S. C., *Rubber World*, 148(5), 1965, 40.
11. TOKITA, N. and PLISKIN, I. *Rubber Chem. Technol.*, 46, 1973, 1166.
12. VAN BUSKIRK, P. R., TURETZKY, S. B. and GUNBERG, P. F. *Rubber Chem. Technol.*, 48, 1975, 577.
13. TURETZKY, S. B., VAN BUSKIRK, P. R. and GUNBERG, P. F. *Rubber Chem. Technol.*, 49, 1976, 1.
14. MILLS, W., YEO, C. D., KAY, P. J. and SMITH, B. R. *Rubber Industry*, 9, 1975, 25.
15. SMITH, B. R. *Rubber Chem. Technol.*, 49, 1976, 278.
16. MILLS, W. and GIURCO, F. *Rubber Chem. Technol.*, 49, 1976, 291.
17. BRYDSON, J. A. *Flow Properties of Polymer Melts*, 2nd edn, Godwin, London, 1981.
18. PALIT, K. and BAR-EL, M. *Eur. Rubber J.*, 162(4), May 1980, 15.
19. BUCKLER, E. J., BRIGGS, G. J., HENDERSON, J. F. and LASIS, E. *Eur. Rubber J.*, 159(7/8), July/August 1977, 21.

20. BRIGGS, G. J., HOLMES, J. M. and WEI, Y. K., *Elastomerics*, **111**(8), August 1979, 30.
21. BUCKLER, E. J., BRIGGS, G. J., DUNN, J. R., LASIS, E. and WEI, Y. K. *Rubber Chem. Technol.*, **51**, 1978, 872.
22. SHELL CHEMICAL CO. Cariflex Technical Bulletin.
23. RODGER, E. R. *Developments in Rubber Technology—1*, A. Whelan and K. S. Lee (Eds.), Applied Science Publishers, London, 1979, Chapter 3.
24. CROUCH, W. and SHORT, J. N. *Int. Rubber Study Group Symposium, Sao Paolo*, 1967, p. 43.
25. WILLIS, J. M. and BARBIN, W. W. *Rubber Age*, **100** (July), 1968, 53.
26. RAILSBACK, H. E. and HAWS, J. R. *Rubber Plast. Age*, **48**, 1967, 1063.
27. COOPER, R. N. and NASH, L. L. *Rubber Age*, **104** (May), 1972, 55.
28. URANECK, C. A. and SHORT, J. N. *J. Appl. Polym. Sci.*, **14**, 1970, 1421.
29. DUCK, E. W., BOWMAN, I. J. W. and WILSON, C. A. Preprints of *Int. Rubber Conf., Brighton*, 1977, II 33-I.
30. HAWS, J. R., NASH, L. L. and WILT, M. S. *Rubber Industry*, **9**, 1975, 107.
31. LIVIGNI, R. A., HARGIS, I. G. and AGGARWAL, S. L. *Int. Rubber Conf., Kiev*, 1978.
32. DAVIS, A. J. *Rubber World*, **181**(6), March 1980, 29.
33. DACKER, K. DE., DUNNOM, D. D. and MCCALL, C. A. *Rubber Age*, **96**, 1969, 53.

Chapter 3

DEVELOPMENTS IN ACRYLONITRILE–BUTADIENE RUBBER (NBR) AND FUTURE PROSPECTS

H. H. BERTRAM

Bayer AG, D-5090 Leverkusen, West Germany

SUMMARY

Nitrile rubber (NBR) has been manufactured on a large scale for more than 40 years and is now the standard elastomer for rubber goods needing resistance to oils, greases, and liquid fuels. In accordance with the desired processing characteristics and properties of the finished goods, NBR is offered in a wide range of grades with several different physical forms. Four decades of experience in the manufacture, compounding, processing, and practical application of NBR have taught manufacturers and processors many ways of optimising certain properties or their combinations in respect of specific uses.

Where certain requirements are concerned, especially resistance to heat and the alcohol-containing liquid fuels of the future, NBR, as known at present, could soon become inadequate. Nevertheless manufacturers and the rubber industry are already working on problems whose solutions will retain for this important synthetic rubber its whole range of applications; in addition they are developing modern processing techniques.

1. INTRODUCTION

Acrylonitrile–butadiene copolymers (NBR) were synthesised for the first time in 1930 by the then I.G. Farbenindustrie in Germany. A synthetic

51

rubber with resistance to oils, greases, and liquid fuels thus became available for the first time. Since 1937, when Buna N was first manufactured on an industrial scale (by I.G. Farbenindustrie), the importance of NBR to synthetic rubber manufacturers, processors, and users has increased continuously.

A comprehensive survey of nitrile rubber was published by Hofmann[1] in 1963, since then Dunn et al.[2] have described further technological progress and prospects of development. In this chapter the present position of NBR is described in general terms and the factors which could affect the importance of NBR in the future are indicated. NBR latex is not considered here.

2. CHEMICAL AND PHYSIOCHEMICAL CHARACTERISATION

2.1. Acrylonitrile Content

The nitrile rubbers that are manufactured on an industrial scale are random copolymers of butadiene and acrylonitrile:

$$+CH_2-CH=CH-CH_2\frac{1}{m} \qquad +CH_2-CH\frac{1}{n}$$
$$\underset{CN}{|}$$

The polar nitrile group confers polar properties on the polymer itself, thus making it resistant to mineral oils, liquid fuels, and technical lubricant greases. The polar nitrile group also influences the glass transition temperature (T_g) and hence the entropy elastic behaviour. The entropy elasticity is connected with the mobility of fairly large chain segments (with a length of about 5 nm). There can thus be no entropy elasticity below the T_g at which the motion ceases. The T_g of loosely crosslinked elastomers is very similar to that of the uncrosslinked rubber. It is determined partly by the configuration of the double bonds of the main chain and, still more, by the number, size, and polarity of the lateral groups.[3]

In the case of NBR the values of T_g and the elasticity behaviour depend mainly on the content of acrylonitrile (Fig. 1).

Figure 2 shows that as the oil swelling increases T_g falls and vice versa. An NBR grade containing 18 % acrylonitrile has the same degree of resistance to oils as a chloroprene rubber (CR) to which only moderate oil resistance can be attributed. At this acrylonitrile content, however, the NBR has a

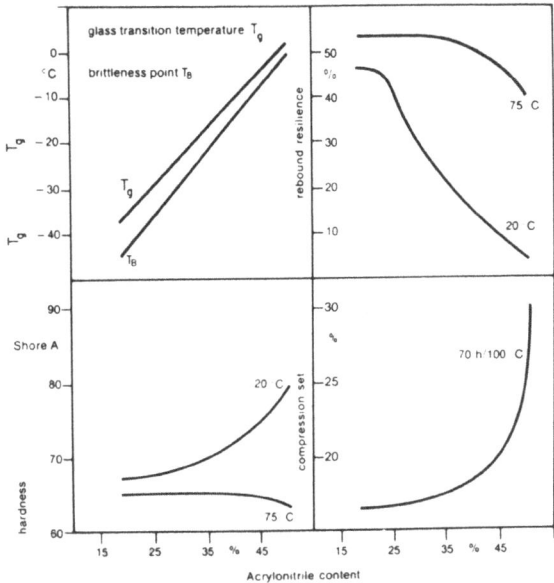

FIG. 1.   Glass transition temperature (T_g), brittleness temperature (T_B), resilience, hardness and compression set versus the acrylonitrile (ACN) content.

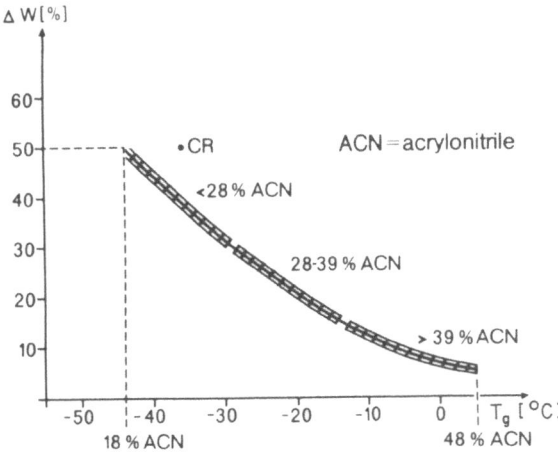

FIG. 2.   Connection between acrylonitrile content, T_g and change of weight after immersion in ASTM oil No. 2.

lower T_g than the equivalent CR. On the other hand the T_g of an NBR grade containing 48% acrylonitrile is very close to room temperature, which means that the applications of the material are greatly restricted.

It follows that suitable compromises must be sought where resistance to oils and liquid fuels is necessary. Hence, as Fig. 2 shows, grades with 28–40% of acrylonitrile are those of the greatest technical importance and the highest and lowest acrylonitrile contents are about 48% and 18%, respectively.

2.2. Copolymerisation and Characterisation[57]

Butadiene and acrylonitrile have different copolymerisation parameters. The azeotropic composition of the monomer mixture is about 38% acrylonitrile (at 25 °C). Where the mixture of monomers contains less than 38% acrylonitrile the copolymers have acrylonitrile contents that are larger than those which would correspond to the acrylonitrile inputs. This results in continuous alteration of the monomer composition and hence in a lack of chemical uniformity. In industrial production additional acrylonitrile is therefore incorporated in a series of stages when definite degrees of polymerisation have been reached.

Landi[83] has confirmed that, as expected, a multiplicity of glass transition temperatures is obtained when normal polymerisation conditions and < 36% acrylonitrile are used.

The macrostructure (i.e. the molecular weight distribution and long-chain branching) and microstructure (i.e. the monomer sequence distribution and the content of butadiene *cis*-1,4; *trans*-1,4; or 1,2 units) of NBR are influenced by the polymerisation conditions. They, in turn, influence the processing characteristics of the compound. For example, the butadiene portion of NBR produced at 28 °C and containing 36% acrylonitrile consists of 10% of 1,2; 12·4% of *cis*-1,4; and 77·6% of *trans*-1,4 linkages. This composition is distributed randomly over the polymer chain.

The characteristics of the molecular weight distribution of NBR have been described in detail by Scholtan *et al.*[84] Nakajima and Harrell[85] have attempted to establish relationships between the viscoelastic behaviour and long-chain branching or gel. However, such relationships are very complex. In the first place the processing behaviour of polymers cannot be characterised by a single quantity. Secondly, the long-chain branching depends on the molecular weight distribution, on the gel and on the short-chain branching, and is therefore a quantity which cannot be considered in isolation.

3. GENERAL CHARACTERISTICS OF NBR

The general characteristics of NBR have been described in the voluminous, informative literature of the well-known NBR producers and will be dealt with here in general terms with a view to making recent developments more readily comprehensible. The general characteristics are summarised in simplified form in Fig. 3.

+ Oil, fuel and grease resistance
+ Good processing characteristics
+ Variety of curing systems
+ Good hot air resistance
   Long term: 90 °C
   40 days   : 120 °C
   3 days   : 150 °C
+ Low permanent set
+ Good abrasion resistance
+ Low gas permeability
− Moderate to good low temperature flexibility
− Moderate ozone resistance (except NBR/PVC)
− Moderate tack
+ Compatibility with polar thermoplastics
   (e.g. PVC, phenolics)

Fig. 3. Basic properties of NBR.

3.1. Swelling Resistance

The good swelling resistance of NBR is, of course, restricted to contact with non-polar or slightly polar substances, such as mineral oils, liquid fuels with low aromatic contents, and greases based on mineral oils. Polar substances, such as esters and ketones, initiate strong interactions and cause severe swelling of NBR; the same applies to aromatic hydrocarbons. NBR is resistant to fuels containing aromatic hydrocarbons, only if the fuels do not contain more than about 50% aromatics. If liquid fuels containing aromatics are additionally modified with alcohols, the resistance to swelling suffers a further drastic reduction, which will be considered later (see Section 7.2).

Where contact with mineral oils is concerned the resistance to swelling does not merely depend on the oil's viscosity and content of paraffinic, naphthenic, and aromatic hydrocarbons. Experiments with technical lubricating and hydraulic oils have shown that the resistance to swelling is also influenced by the types and amounts of additional additives, by the temperature and by the exposure period. These factors have been investigated systematically by Bertram and Brandt.[4] It follows that a vulcanisate which has been found to resist certain technical lubricants will not necessarily resist others.

3.2. Ageing Behaviour

The resistance of NBR to ageing in hot air is superior to that of natural rubber and polyisoprene. As will be seen (Section 6), much progress has been made in this respect in recent years by improving the choice of compounding ingredients and their ratios. Accordingly it is now possible to make NBR vulcanisates 'permanently resistant' to hot air at about 90 °C. This means that they can be exposed continuously to air at this temperature for approximately 12 months. At 120 °C a service life of about 40 days can be expected; at 150 °C it is likely to be about three days.

3.3. Resistance to Ozone

The ozone resistance of unprotected NBR is comparable with that of other diene rubbers. Unlike that of natural rubber and styrene–butadiene rubber, however, it cannot be greatly improved by including paraffinic hydrocarbons and antiozonants in the compound. In cases where increased thermoplasticity and reduced low-temperature flexibility are acceptable this difficulty is often overcome by blending PVC with the NBR (see Section 5.3.1).

3.4. Applications

3.4.1. Vulcanisate Property Requirements

In the great majority of applications NBR is used because of its resistance to oils, liquid fuels, and greases. In most cases the vulcanisate must also have

1. A specific combination of definite mechanical properties.
2. A specific degree of low-temperature flexibility (as determined by a defined method).
3. A specific degree of resistance to ageing, which will be related to the envisaged service temperature.

Of all synthetic rubbers, NBR is therefore the one with the most complicated finished article specifications and the one for which the largest number of such specifications exists. It should be pointed out in this connection that the difficulty lies not in obtaining a particular characteristic but in obtaining a particular combination of several characteristics or properties. Furthermore, even where a finished article specification has

TABLE 1

NBR USED IN RUBBER GOODS (WEST GERMANY)

Rubber goods	%
Moulded technical goods	38
Hoses and tubes	26
Rubber rollers	15
Brake linings and clutches	8
Cellular goods	4
Coated fabrics	3
Special footwear solings	2
Cables and wires	0·5
Linings	0·5
Others	3
Total	100

been met, one cannot be certain that the rubber article will function satisfactorily until a practical trial has been completed.

3.4.2. Areas of Use

The applications of NBR depend on the structure of industry and therefore differ to some extent from country to country. The applications of NBR in the Federal Republic of Germany are estimated to be as shown in Tables 1 and 2.

TABLE 2

INDUSTRIES (WEST GERMANY) USING NBR

Type of industry	%
Motor vehicle	38
Machinery	21
Food	12
Textile	7
Printing	5
Footwear	2
Electrical goods	3
Construction	4
Iron and steel	1
Mining	1
Chemical	0·5
Others	5·5
Total	100

In 1979 the NBR consumption (without latex) of the non-communist countries was about 203 000 t. Further increases will be closely dependent on general economic growth in those industries in which NBR goods are principally used. Consumption is expected to increase by 3·8 % per annum in terms of volume. Above-average growth is expected in cellular thermal insulations (NBR/PVC blends) and high and very high pressure hoses.

4. RECENTLY INTRODUCED GRADES

NBR solid rubber is made by 20 producers, who have altogether 324 grades[5] which differ in their acrylonitrile content, Mooney viscosity, stabiliser, and polymerisation temperature. Commercial products cover an acrylonitrile range of 18–50 % and a Mooney viscosity $ML(1 + 4)/100\,°C$ range of 20–140. The commercially most important grades have acrylonitrile contents of 28–40 % and Mooney viscosities of 25–100.

The processing properties of some grades are improved by precrosslinking with small amounts of divinyl benzene or difunctional acrylates at the polymerisation stage. The following brief descriptions will be limited to recently introduced types of NBR.

4.1. Crumbed Rubbers
Some grades are supplied not only as bales but also in a mechanically granulated form known as crumb. The crumbs are treated with a separating agent to prevent their agglomeration. Crumb is used particularly in the manufacture of brake linings and rubber-bound asbestos seals, where the rubber has to be dissolved.

4.2. Powdered Rubbers
Powdered rubbers can be produced by mechanical grinding, spray drying, or special coagulation processes. This form of NBR allows new and cost-saving compounding techniques to be used (see Section 7.4); its use also speeds up compounding by the traditional methods and facilitates the handling of NBR-modified thermoplastics (e.g. PVC).

4.3. Carbon Black Masterbatches
As in the case of SBR,[6] carbon black masterbatches can be produced at the polymerisation stage. Although they offer the advantage of reduced compounding times[7,8] they have failed to become widely established

because NBR is used in a very great variety of recipes. This means that the acrylonitrile content and viscosity of the NBR and the type and amount of carbon black must be variable. NBR-black masterbatches are also available in particulate forms to which plasticisers can be added by a dry blend method.

4.4. Plasticiser-modified NBR Grades

Unlike the addition of mineral oils to SBR, which results in what are known as oil-extended grades and is intended to reduce prices, the addition of ester and ether plasticisers to NBR at the polymerisation stage is done in order to reduce the long compounding times of baled rubber. This is particularly true when very high proportions of plasticiser are incorporated. Nevertheless, the resulting materials are only important in the production of soft vulcanisates—printing rolls, for example. Grades containing plasticisers are also offered as NBR/PVC blends.

4.5. Carboxylated NBR (XNBR)

NBR with randomly distributed carboxyl groups is obtained by the terpolymerisation of butadiene, acrylonitrile, and methacrylic or acrylic acid. Compared with ordinary NBR, these materials have better abrasion resistance and higher tensile strength and tear resistance values.[10]

The advantages just mentioned are attributed to the ionic crosslinking of the carboxyl groups with zinc oxide. As the ionic crosslinks have little heat stability the vulcanisates have only moderately good compression set behaviour at elevated temperatures. The various ways in which XNBR can be crosslinked have been reviewed by Brown.[11] For a number of years the commercial importance of XNBR was limited by the scorching which tends to accompany the crosslinking action of the ZnO. One solution to the problem is that provided by Hallenbeck,[12] who coated the particles of the ZnO with zinc sulphide or zinc phosphate.

An alternative is to crosslink the polymer with zinc peroxide.[12,13,13a,71] It is uncertain whether either of these solutions will increase the importance of XNBR decisively.

4.6. Network-bound Antioxidant NBR Grades

Over the last 10 years the demands made on rubber goods in respect of resistance to ageing in hot air and hot lubricating oils have increased continuously. Where NBR is concerned, much can be done by optimising the whole recipe, as will be seen in Section 6. There are, however, several approaches by which the resistance to ageing is improved by modifying the polymer itself.

To prevent losses of antioxidants through volatilisation or extraction, Kline and Miller[14] have proposed the use of antioxidants which co-polymerise with butadiene and styrene or butadiene and acrylonitrile.[15–18] They draw particular attention to N-(4-anilinophenyl) methacrylic amide, which is used for a number of special grades.

Another approach to the same goal is described by Scott and co-workers[19–21] who have proposed the use of nitrones and the grafting[22] of suitable antioxidants on to the polymer chain.

Weinstein,[23,24] on the other hand, has attempted to attach antioxidants based on symmetrical dithiophenols or phenol- or amine-substituted sulphur compounds to the polymer chain during or after the polymerisation. The purpose of all such developments is to optimise the ageing resistance of standard NBR by adapting the compound to the specific requirements (see Section 6).

As this field of technology is very largely in a state of flux, a definitive assessment cannot yet be attempted.

4.7. Hydrogenation of NBR

The double bonds of NBR have been selectively hydrogenated for the same purpose, i.e. in order to improve the resistance of the vulcanisates to ageing in oils and hot air. Pyridine–cobalt complexes[25] and complexes of rhodium, ruthenium, and iridium[26–29] have been described as hydrogenation catalysts. In some cases the hydrogenation is incomplete.[28] The main obstacle to homogeneous catalytic hydrogenation of olefinic structures is the difficulty in obtaining adequate selectivity, but there are special catalysts based on transition metals which are almost entirely satisfactory in this respect.[29]

The almost completely hydrogenated polymers are crosslinked with peroxides. They have very good mechanical properties (tensile strength and resistance to abrasion), an exceptionally low brittleness temperature, excellent ozone resistance, and considerably improved resistance to ageing in hot air and hot industrial lubricants. These technically interesting new polymers, which can no longer be looked upon as NBR in the traditional sense, are not yet commercially available.

4.8. Epoxy-modified NBR

With 0·07 mol of epoxy groups Edwards and Sato[30] obtained modified NBR materials whose interactions with the silanol groups of silica fillers are similar to those which take place between ordinary rubbers and, for example, mercaptopropyltrialkoxysilanes[31–33] or bis(triethoxydisilylpropyl) tetrasulphides.[34] These products are not yet commercially available.

4.9. Liquid NBR

Liquid forms of NBR with or without randomly distributed functional groups, such as —COOH, have been known and used for a long time. In the rubber industry they are added to NBR compounds that would otherwise be difficult to process because the compounds possess a high viscosity. As they, too, are subjected to the crosslinking reactions, they also serve as nonextractable or not easily extractable plasticisers.

In addition, rubber technologists have been trying for a long time to imitate the casting of urethane elastomers by applying the technique to low-molecular analogues of the well-known rubbers. The object is shaped while the viscosity of the compound is still low, after which the polymer network is built up. The compound can only be processed in the liquid state, however, if the average molecular weight of the prepolymer is not too high (M_n must be 4000–6000). Nevertheless, optimum properties are only obtained when the distance between two crosslink sites is greater than the chain lengths of the prepolymers. Chain extension is therefore necessary.

For that purpose every prepolymer molecule must have two reactive groups and these must, as far as possible, be situated at the chain ends (telechelic polymers). Every molecule must be exactly difunctional and the reactivity of the ensuing crosslinking step must agree with the preceding chain extension. Liquid NBR is produced with carboxyl, hydroxyl, mercaptan, amine, vinyl, or halogen terminal groups.

Drake and McCarty[35] have described the potential applications of the materials and the various ways in which these can be crosslinked. For several reasons, however, the manufacture of rubber goods based on NBR telechelics has failed to establish itself. One reason is that the available products are not exactly difunctional. Another reason is that reinforcing fillers make it difficult or impossible to process liquid NBR but are, however, necessary. In the absence of these fillers the mechanical properties of the vulcanisates are not comparable with those of ordinary NBR vulcanisates (it was therefore necessary to develop a new processing technique[36]). Finally, the prepolymers are still so expensive that the finished products have no cost advantage.

A proposal which could lead to technically satisfactory vulcanisates has been discussed by Kalfogolou and Williams[37] and described in detail by Siebert.[38] A 50:50 blend of carboxyl-terminated NBR with epoxy resin and

4,4'-isopropylidene phenol has been found to give a castable composition which, having been crosslinked, has good mechanical properties and resistance to ageing.

Until now the industrial use of these reactive NBR liquids has been restricted to the modification of thermosetting resins. For example, liquid carboxyl NBR increases the rupture energy of epoxy resins,[39] while vinyl NBR increases that of unsaturated polyester resins.[40]

4.10. Alternating NBR

Furukawa and co-workers[41,42] have developed catalysts from alkyl aluminium halides and transition metal halides, such as VCl_3 or $TiCl_4$, which permit the production of equimolar butadiene–acrylonitrile copolymers (containing 48 % acrylonitrile).

The acrylonitrile must be complexed with the catalyst before the butadiene is added and such a copolymer has a degree of alternation of 94–97 %.[43,44] The butadiene has mainly *trans*-1,4 structure as compared with randomly structured copolymers with comparable acrylonitrile contents. Alternating copolymers have increased green strength and, when vulcanised with the same crosslinking system, higher tensile strength values, a fact which Furukawa *et al.*[45] have attributed to overcuring of the emulsion polymer. As the structure of alternating NBR is regular, the increased green strength is a consequence of the strain crystallisation.

Until recently alternating NBR has been of academic interest only, as it is expensive to produce and the market potential of NBR containing 48 % of acrylonitrile is limited.

4.11. Thermoplastic NBR

Although thermoplastic elastomers with other chemical compositions have already acquired considerable importance, no thermoplastic NBR is yet available. The state of development has been reviewed by Dunn *et al.*[2] An interesting method of obtaining thermally or mechanically reversible crosslink sites has been proposed by Lasis *et al.*[46] and is described in several previous patents.[47–50]

According to this method unstable crosslink sites are produced by reactions between tertiary amines and organic polyhalogen compounds. The tertiary amine is introduced into the polymerisation batch as a comonomer by acrylates or methacrylates having the formula

$$H_2C = C - C - O(X)N(CH_3)_2$$

with R above the central carbon and O above the carbonyl carbon.

where R = H or CH$_3$ and X = aliphatic hydrocarbon or secondary or tertiary amine-substituted aliphatic hydrocarbon with two to four carbon atoms.

The polymers have increased green strength and are said to permit the reuse of scrap formed in the moulding process. It must be assumed that 'unstable systems' of this kind do not share the good processability that is exhibited by conventional systems which can be processed by techniques characteristic of thermoplastics.[2]

The development of a technically satisfactory thermoplastic NBR is eagerly awaited.

4.12. Acrylonitrile/Isoprene Rubber (NIR)[51–55]

These copolymers are not, strictly speaking, NBR. However, they are very closely related and will therefore be briefly considered. Compared with NBR, NIR copolymers are more easily depolymerised, are easier to extrude and they have higher tack. Shapiro *et al.*[54] found by the method of ^{12}C NMR spectroscopy that NIR contains a high proportion of triads. Gatti and Carbonaro[55] have carried out detailed investigations by the same method.

In consequence of strain crystallisation even vulcanisates without fillers have high tensile strength. The elasticity and low-temperature flexibility values are considerably inferior to those of NBR. NIR has been commercially available for several years but the various grades have not yet acquired much economic importance.

Copolymers produced with complex catalysts and having high degrees of alternation have been described[56] in addition to randomly structured emulsion polymers.

5. MODIFICATION BY BLENDING WITH OTHER POLYMERS

Blends of NBR with other rubbers or with thermoplastic or thermosetting resins have been known for several decades. Modification by blending is undertaken to reduce costs, to improve the rheological properties of material, or to confer specific physical or chemical properties to the end product.

5.1. General Properties

The fundamental characteristics of the end product are determined chiefly by the polymer which is present in the larger proportion of the blend. Thus a blend consisting mainly of NBR has, on the whole, the typical properties of NBR. A blend consisting mainly of a thermoplastic and containing only a small amount of NBR will have the fundamental characteristics of the thermoplastic, though certain of its properties (e.g. toughness) may have been substantially improved. The role of elastomers in rubber-modified plastics has been discussed at length by Dinges.[58] Only polymer blends consisting mainly of rubber will be considered here.

The properties of all polymer blends depend on the effectiveness of the physical blending process; they also depend on whether the polymers are compatible or form a heterogeneous mixture, and therefore—in the latter case—on the morphology of the blend.[59,60] Also of importance in the case of elastomer blends is the chemical process of crosslinking. This is very complex in heterogeneous blends because (i) different rubbers have different rates of crosslinking; (ii) the crosslinking system and plasticisers have different degrees of solubility in the various rubber phases; and (iii) the interactions between the rubber and filler differ according to the rubber concerned.

5.2. Blends of NBR with Other Rubbers

5.2.1. NBR/NR and NBR/SBR Blends[61]

These blends have been used in the rubber industry for several decades— NBR/SBR blends for reasons of cost or to meet certain swelling resistance requirements and NBR/NR blends because they have more tack than NBR alone.

5.2.2. NBR/CR Blends[61]

NBR/CR blends are more resistant to ozone and fatigue than NBR alone, yet they retain good resistance to oils. They are, however, less resistant to ozone than NBR/PVC and NBR/EPDM blends.

5.2.3. NBR/EPDM Blends[61-63]

These blends, similarly to those consisting of NBR and PVC, are intended to combine the ozone resistance of EPDM with the oil resistance of NBR. NBR/EPDM blends lack a homogeneous phase because the difference in the polarity of the two materials is excessive. The heterogeneity of the blends can be seen in Fig. 4. With a view to obtaining a finer dispersion of the EPDM particles in the NBR two manufacturers produce NBR/EPDM blends by mixing latices. It is not yet possible to assess the suitability of the resulting materials for their intended purposes.

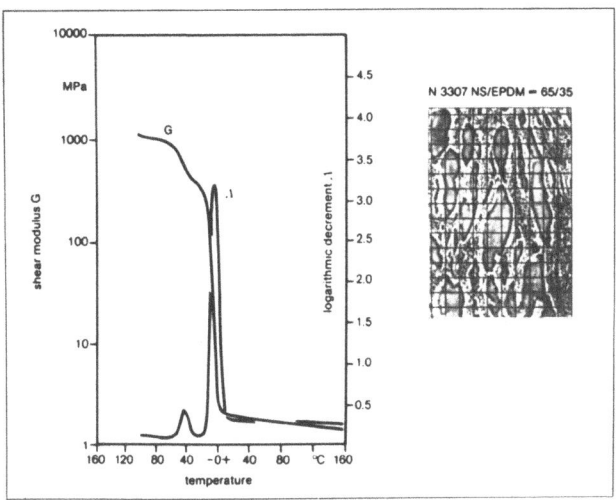

FIG. 4. Torsional oscillation test according to DIN 53520 on blends of NBR (34 % acrylonitrile) with EPDM and their dispersion photographs. Blend ratio 65 NBR/35 EPDM. Logarithmic decrement and shear modulus as functions of temperature.

For good ozone resistance the EPDM must be dispersed as finely as possible. The blending conditions are therefore decisively important. In this connection Mitchell[62] recommends that carbon black, plasticisers, and activators be added exclusively to the EPDM. Other authors[63] recommend the addition of chemicals which have homogenising effects. Good resistance to ozone can be obtained with blends containing 30–35 % EPDM but in many cases, where resistance to swelling is important, the proportion of EPDM cannot exceed 20 %. Because of this waxes and antiozonants have to be included (Fig. 5).[61]

The problems that arise from the different rates of crosslinking and accelerator solubilities have been investigated by many authors.[61,64–67] The classical thiuram disulphide vulcanisation systems, possibly supplemented by sulphur donors, give the best overall results.[61] Long-chain alkyl dithiocarbamates[67] are also recommended as special accelerators. Peroxide crosslinking gives good ozone resistance (Fig. 5) but the overall property pattern is only moderately good because the NBR is overcured and the EPDM undercured.[61] In principle, NBR/EPDM blends can be used to combine the swelling resistance of NBR with the ozone resistance of EPDM. But problems, especially in connection with goods subjected to

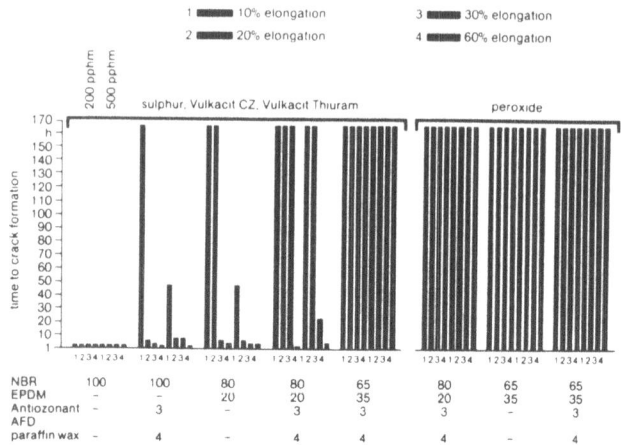

FIG. 5.   Ozone resistance of blends of NBR (34 % acrylonitrile) with EPDM.

dynamic stresses,[61] arise from the heterogeneity of the blends and from the fact that their components have different rates of crosslinking.

5.2.4. NBR/BR Blends[61]

Here again the differences in polarity make the polymer mixture heterogeneous. The inclusion of 10 % BR considerably improves the performance in the cold bending test (according to ASTM-D 736-57 T). At approximately 20 % BR the brittleness temperature is reduced considerably without the dynamic glass transition temperature being altered. Other effects of BR addition are an improvement of the flow behaviour in injection moulding and a reduction of the energy needed for extrusion. Certain adverse effects on the physical properties of the vulcanisates have to be accepted.

5.2.5. NBR/EVAC Blends[61]

Vulcanisates which withstand oils and ozone can be produced with blends of NBR with about 30 % ethylene vinyl acetate rubber (vinyl acetate content > 40 %). Compared with NBR/PVC blends such blends have the advantage of being more resistant to ageing. Thiuram disulphide or peroxide can be used for crosslinking.[61] The physical properties are inferior to those of all-NBR vulcanisates. Small additions of NBR to EVAC compounds improve the mould release and the peelability of electric wires.

5.2.6. NBR/CIIR Blends[61]

These blends, too, are heterogeneous. Good resistance to ozone is obtained only with large proportions of CIIR, but these impair other properties of the vulcanisates.

5.2.7. Blends of NBR with Other Special Rubbers
Propylene oxide rubber (PO) shows depolymerisation phenomena at high ageing temperatures, whereas NBR undergoes further crosslinking and becomes harder. The attempts that have been made to improve the ageing resistance of rubber goods by blending these two polymers with a view to combining their opposite tendencies have not yet been successful.[61]

Similarly, in cases where NBR has remained an important constituent of the polymer mixture it has not proved possible to increase the resistance of NBR vulcanisates to ageing by blending with epichlorohydrin (ECO)[61] or fluoro (FPM) rubbers.

5.2.8. Blends of NBR with NBR
Blends of standard NBR grades with precrosslinked NBR have long been used to obtain definite processing characteristics. No further reference will be made to them here.

Blends of standard NBR grades with one another will however be briefly considered.

Grades whose acrylonitrile contents differ widely do not give homogeneous mixtures.[68] As Fig. 6 shows, the mixture is still homogeneous when the acrylonitrile contents of the NBR are 18% and 28%. But mixtures of grades containing 18% and 34% acrylonitrile have two maxima in the torsional oscillation test, which indicates that they are heterogeneous.

In the case of heterogeneous blends the low-temperature service temperature is not determined exclusively by the blending component which has the lower glass transition point; it also depends on the blending ratio and degree of incompatibility.

As, in addition, the swelling resistance of heterogeneous NBR blends is slightly inferior to that of homogeneous blends, their use provides almost no opportunity to improve the relationship between low-temperature flexibility and resistance to swelling.

5.3. Blends of NBR with Plastics
Although small amounts of NBR are added to thermoplastic or thermosetting resins to improve certain properties of these, the following remarks are concerned mainly with blends in which the rubber predominates and which are processed and used similarly to rubber.

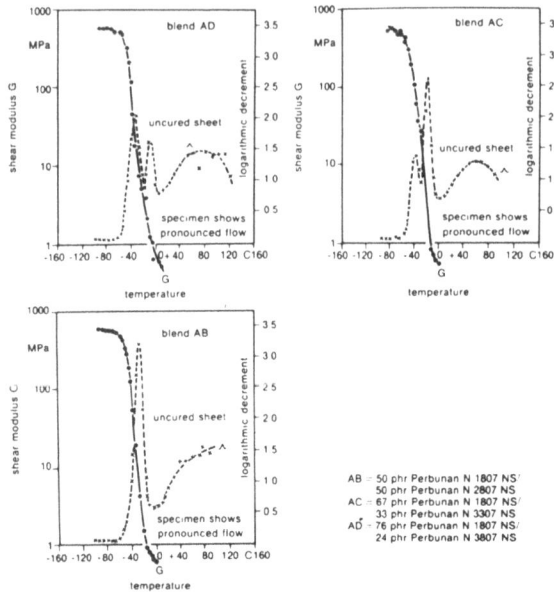

FIG. 6.   Torsional oscillation test with blends of NBR having different acrylonitrile contents; logarithmic decrement and shear modulus versus temperature.

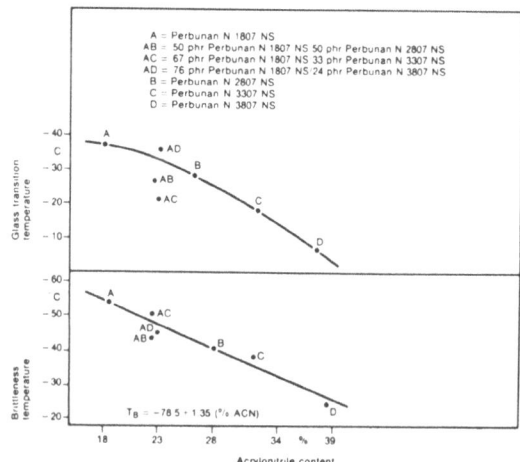

FIG. 7.   Blends of Perbunan N grades with different acrylonitrile contents. Glass transition temperature and brittleness temperature versus the acrylonitrile content.

Only a few of the many possible blends will be mentioned. As little work was done in this field until recently, further interesting developments can be anticipated.

5.3.1. NBR/PVC Blends

Blends of NBR with PVC were described by Konrad[9] as early as 1936. The main advantage is that they combine the ozone resistance of PVC with the crosslinkability and oil resistance of NBR. Since 1962 they have enjoyed extensive commercial importance.

5.3.1.1. Production methods. Abrams[69] pointed to the necessity of fluxing the NBR and PVC jointly so that the vulcanisates would have the best possible mechanical properties and resistance to ozone. The fluxing or mixing must be performed at a temperature which is above the glass transition temperature of the PVC. Although this can be performed on heated rubber-processing machines, fluxed blends are offered by most NBR producers. These are produced by either of two methods. In one method the solid polymers are mixed mechanically at a temperature exceeding the glass transition temperature of the PVC. In the other method a PVC dispersion is mixed with NBR latex and then the polymer mix is coagulated, washed, and fluxed.

5.3.1.2. Nature of the blend. Figure 8 shows that blends of NBR with PVC have a continuous property spectrum, i.e. one which entirely reflects the ratios of the two materials. At PVC contents of up to about 50 % the

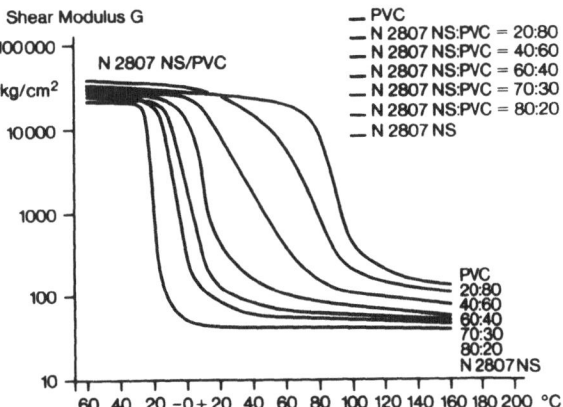

FIG. 8.   Free torsional oscillation test. Shear modulus G versus temperature at different blend ratios of NBR (28 % acrylonitrile) and PVC (crosslinked compounds).

properties of the NBR predominate. These blends are processed by the methods of the rubber industry and are therefore vulcanised. At PVC contents of more than 50% the properties of the thermoplastic are predominant. These blends are processed and used similarly to thermoplastics; the NBR is no longer crosslinked and it serves as a polymeric plasticiser or impact modifier.[72-75] Where such blends are concerned the acrylonitrile content, degree of branching or precrosslinking, and particle size of the NBR are decisively important. The rest of this section will be devoted only to NBR/PVC blends in which the NBR predominates.

5.3.1.3. Degree of homogeneity. The opinions of experts still differ as to whether NBR and PVC form a homogeneous polymer phase. It is accepted however, that NBR containing less than 23–24% of acrylonitrile fails to give a homogeneous polymer phase with PVC, whereas homogeneity has been reported for blends whose NBR contains 34–40% acrylonitrile.[76,78,82]

However, Jordan et al.[77] found inhomogeneity by the DSC method; they attributed their finding to the ability of this method to determine considerably smaller particles than those determined by mechanical measurement methods.

5.3.1.4. Ozone resistance. The main technical advantage of NBR/PVC blends over NBR is that they have excellent resistance to ozone. For this they must contain not less than about 30% PVC.

Khanin et al.[79] attribute the improvement to a blocking of the double bonds of the NBR by the PVC segments. In common with other authors, they draw attention to the influence exerted by the production conditions. That is why the experts differ as to whether a latex blend or a well fluxed solid polymer blend gives better results.[80,81] The main considerations are that effective PVC stabilisers should be present during the fluxing process and that this should be accompanied by mechanical mixing, performed at a sufficiently high temperature (160–170 °C), and continued for a sufficient period of time. The influences exerted by the acrylonitrile content and Mooney viscosity of the NBR, by the molecular weight of the PVC, and by the blend ratio have been investigated on several occasions.[70,78] For favourable compromises in respect of the processing and vulcanisate properties, PVC with a Fikentscher viscosity of 60–70 and NBR with a Mooney viscosity of 30–45 and an acrylonitrile content of 28–34% are preferred.

5.3.1.5. Application of NBR/PVC blends. NBR/PVC blends are of considerable technical importance in the manufacture of cellular thermal insulations, fire and irrigation hoses, fuel hose covers, cable jackets, spinning aprons, soft rollers, and as modified PVC. Their use is restricted by deficiencies in low-temperature behaviour, compression set behaviour, resilience and resistance to ageing.

5.3.2. Blends of NBR with Polyvinyl Acetate (PVAC) and its Copolymers

Stollfuss[87] investigated blends of NBR with polyvinyl acetate (PVAC) and its copolymers. The object was to obtain better low-temperature flexibility, compression set behaviour, and ageing resistance than that possible with NBR/PVC blends. Although the solubility parameters are similar in the two cases and the formation of a homogeneous phase was expected, only two-phase systems were found, except in the case of a vinyl acetate/maleic acid ester copolymer. All the mechanical properties, and also the resistance to ozone, were inferior to those of the NBR/PVC blends. Only the resistance to ageing was improved.

5.3.3. Blends of NBR with Cellulose Esters[87]

According to the solubility parameters blends of NBR with cellulose acetate, propionate, or acetate butyrate should give a homogeneous phase. Owing to the formation of hydrogen-bridge bonds, however, the theory fails; the polymer blends consist exclusively of two phases and they have no resistance to ozone, even when the fluxing has been carried out at temperatures of around 180 °C.

The vulcanisates combine great hardness with good resilience, but the compounds are exceedingly viscous and therefore difficult to process.

5.3.4. Blends of NBR with Polyamides (PA)[86]

Despite the similarity of the solubility parameters, pronounced crystallisation and hydrogen-bridge bond formation by the polyamides (PA) prevent NBR/PA blends (6-PA, 6, 6-PA) from forming a homogeneous phase. Tolstukhina and Kolesnikova[86] found that the benzene resistance of gelled blends was considerably improved, but at the expense of the resistance to abrasion and elongation at break. It is assumed that the NBR is crosslinked by the polyamide. Until recently the need for a very high fluxing temperature and the fact that the material is very viscous and therefore difficult to process have been disadvantageous.

5.3.5. Blends of NBR with Phenolic Resins

There are two reasons for producing these blends. Firstly, NBR can be crosslinked with reactive phenolic resins in the presence of activators— organic acids, such as fumaric and pyromellitic acid, or halogen-containing compounds. Secondly, novolacs without reactive groups can be added to NBR as reinforcing resins and hardening brought about by adding hexamethylene tetramine.

In the second of these cases the resins act as reinforcing fillers, giving the goods high tensile strength and hardness. The reinforcing effect depends on the types of resin and NBR and on the dosage of hexamethylene tetramine.[88] Kosfeld and Borowitz[89] have confirmed that the resulting polymer mixtures are homogeneous.

NBR/phenolic resin blends have been produced industrially for many years, either for NBR rubber goods that must be hard and resistant to swelling or as a means of raising the impact resistance of phenolic resins.

6. COMPOUNDING TO OPTIMISE SELECTED PROPERTIES

The experience of NBR compounding which has been gained by manufacturers and processors of this material extends over several decades. The following section is therefore confined to a few recent studies which have been aimed at optimising certain properties of the compounds and vulcanisates.

6.1. Rheology of Compounds

As the compounding of NBR generally raises no difficulties there are very few systematic studies of it. A recent investigation[90] states that the dispersion of compounding ingredients is best when moderate compounding temperatures (65–100 °C) and high shear rates are used.

The rheological behaviour of NBR and of NBR compounds loaded with carbon black is discussed by Nakajima and Collins.[91,92] According to these authors the viscoelastic behaviour of amorphous rubbers can be described by a master curve (temperature/time, pressure/time, stress/time) which permits conclusions as to the processing behaviour of these rubbers on mixing mills. Bittel[93] has reported that the Monsanto processability tester is able to distinguish between NBR grades which are equal in Mooney viscosity, but have different shrinkage tendencies, and also that it is able to indicate the state of mix.

On the whole, however, no simple laboratory method that enables the practical processing behaviour of NBR compounds to be predicted has yet been published.

6.2. Vulcanisation

In contrast to other synthetic rubbers NBR can be vulcanised with a wide range of crosslinking systems and these can therefore be chosen to suit the envisaged processing method and to favour the properties desired in the end product. The range of up-to-date crosslinking systems for NBR extends from those that are effective at room temperature (these are used for rubber linings, etc.) to those which at high temperatures give very short, and hence, economical vulcanisation times. NBR can also be vulcanised with crosslinking systems which, under the legislation of various countries, are permitted to be used in the manufacture of food-contact rubber goods. Among the methods that have been applied to NBR most recently are high energy radiation curing and curing initiated by ultraviolet light. The various vulcanisation systems for NBR are described in the printed literature of well-known NBR manufacturers.[94,95] Thus only three recent developments will be described.

6.2.1. Injection Moulding

An important requirement of the crosslinking systems for injection moulding is high processing safety because this enables the injection temperature to be raised, thus improving the flow. In addition the vulcanisation time must be short so that economic production cycles can be achieved.

Until now the use of crosslinking systems based on thiuram disulphide for compounds which reach high temperatures has been difficult owing to the inadequacy of the scorch times. However, Leibbrandt[96] has drawn attention to the development of systems comprising tetramethyl thiuram disulphide (or preferably dimethyl diphenyl thiuram disulphide), N-morpholinyl-2-benzothiazyl sulphenamide, a special retarder (Vulkalent E), and 0·5–0·8 phr of sulphur which, especially in the case of NBR, combine high processing safety with short vulcanisation times. The thiuram disulphide can be replaced by other sulphur donors (see Table 3 and Fig. 9).

TABLE 3

Vulcanisation system	1	2	3	4
Tetramethyl thiuram disulphide	2·5	1·5	—	—
N-cyclohexyl-2-benzothiazyl sulphenamide	2·0	—	—	—
Dimethyl diphenyl thiuram disulphide	—	—	2·0	—
Caprolactam disulphide	—	—	—	1·5
N-morpholinyl-2-benzothiazyl sulphenamide	—	2·0	2·0	2·0
Vulkalent E	—	1·0	1·0	1·0
Pasted insoluble sulphur (75 %)	0·27	0·54	0·8	0·8

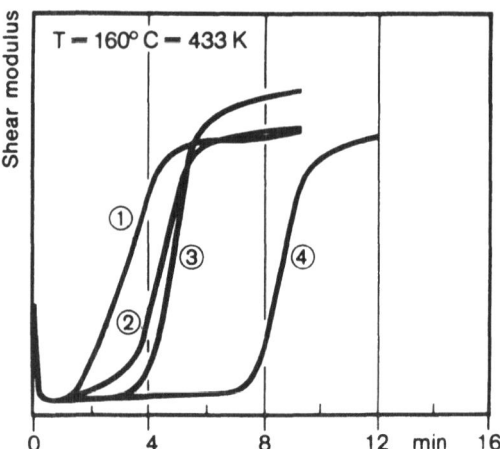

FIG. 9. Curing curves of injection moulding systems.

These systems and slight modifications of them have proved highly suitable for NBR.

6.2.2. Crosslinking Systems for Room Temperature

In the rubber-lining of pipes and large vessels autoclave vulcanisation is superfluous if vulcanisation systems are available which crosslink the compound at the ambient temperature over a period of several weeks and which have just sufficient processing safety for this application.

In addition to System A, (2·4 phr of sulphur, 1·0 phr of mercaptobenzothiazole, and 0·5 phr of diphenyl guanidine), which has been used successfully for many years, the following ones, which have similar or greater degrees of efficiency, have been developed:

System B = 2·4 phr sulphur + 0·5 phr mercaptobenzothiazole + 1·0 phr zinc-N-ethyl-phenyl-dithiocarbamate.

System C = 2·4 phr sulphur + 0·5 phr zinc-N-ethyl-phenyl-dithiocarbamate + 0·5 phr cyclohexyl amine + 0·5 phr N-cyclohexyl-2-benzothiazyl sulphenamide.

System D = 2·4 phr sulphur + 1·0 phr diphenyl guanidine + 1·0 phr mercaptobenzothiazyl disulphide (2-benzthiazyl disulphide).

6.2.3. Crosslinking with the Aid of Ultraviolet Radiation

The crosslinking of acrylic acid esters in the presence of photo-initiators and under the simultaneous action of ultraviolet light has been known for a long time. It has been applied also to unsaturated polyester (UP) resins, polyurethanes, and polyamides.

The principle is also applicable to NBR, which is mixed with diacrylic or triacrylic acid esters and a photo-initiator and then exposed to ultraviolet radiation. Benzophenone and diacetyl are among the photo-initiators used. Printing plates and rollers and similar goods can be produced by localising the irradiation and removing the uncrosslinked polymer with a solvent.[97–99] Until recently the method has been applied to NBR to a limited extent only.

6.3. Heat Resistance

A great deal of literature has been published on the opportunities for improving the resistance of NBR goods to ageing in hot air and hot lubricating oils.[2,14–24,100–109] The reason why so much work has been done in this sector is that the motor vehicle industry, in particular, has been demanding increasingly high degrees of resistance to ageing, especially over long periods.

Lee and Morell[100] found by means of stress relaxation measurements that NBR in the absence of air is thermally stable at temperatures up to 150 °C. This was subsequently confirmed by Pfisterer and Dunn.[105] The use of the Cadmate system,[109] which represented a substantial advance, is restricted by considerations of toxicology.

Several NBR manufacturers[14–18] have endeavoured to counteract the volatility and extractability of antioxidants by combining these with the polymer chain (see Section 4.6). Other work has been directed at optimising the properties of vulcanisates based on conventional NBR grades through improved compounding.[102–107,110] The outcome is that the resistance to ageing of these vulcanisates is best when they contain mineral fillers, silanes, plasticisers of low volatility, a thiuram disulphide vulcanisation system (or possibly a vulcanisation system based on another sulphur donor), zinc mercaptobenzimidazole, and an amine as an antioxidant.[104,110] The exclusion of acidic fillers is important.[102,103] In hot air at 140 °C resistance to ageing for about three weeks has been obtained.

As Pfisterer and Dunn[105] have confirmed, this result is equal to that obtained with NBR grades containing antioxidants which are bound to the polymer chain. Nevertheless, difficulties are still posed by the ageing of NBR parts in certain technical lubricating oils[4,108] whose additives render the vulcanisates useless even at only moderately elevated temperatures. Apart from the use of hydrogenated NBR[29] no generally effective remedy is known.

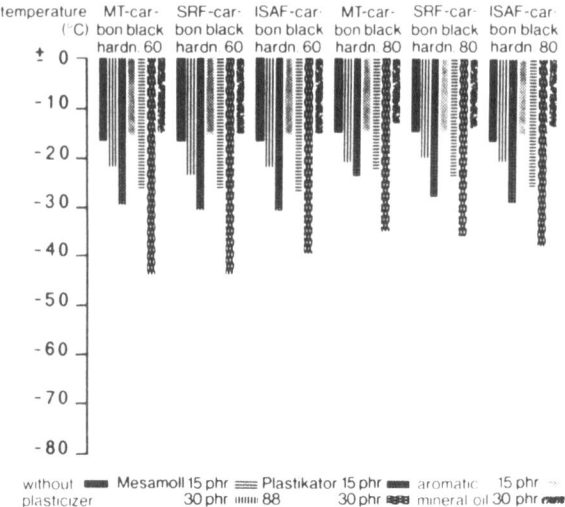

FIG. 10. Second order transition temperature according to DIN 53513 (dynamic test). NBR with 34% acrylonitrile.

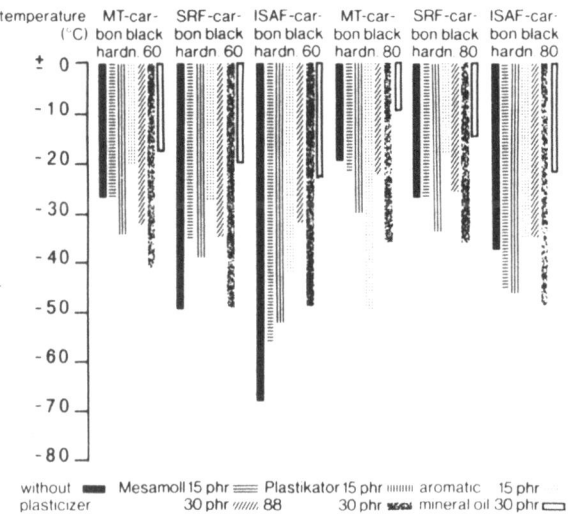

FIG. 11. Brittleness temperature according to DIN 53546. NBR with 34% acrylonitrile.

6.4. Low-temperature Behaviour

Generally speaking, the desired low-temperature behaviour of NBR vulcanisates is not a property that can be considered in isolation from others as every desired property must also be taken into consideration. In this connection attention is drawn to a publication by Engelmann,[111] who pointed out that the various low-temperature test methods give different results and suggested ways in which the low-temperature behaviour, as determined by each individual method, could be optimised.

Engelmann's principal findings are presented in Figs 10 and 11. With test methods which respond to the material's entropy elasticity, the acrylonitrile content of the NBR and the type and amount of plasticiser exert a decisive influence. With test methods which determine toughness and brittleness characteristics at low temperatures, the rubber–filler interactions are important; where polymer blends (e.g. NBR/BR) are concerned, the morphology, too, plays a decisive part:

For optimum low-temperature behaviour (according to the T-50 Test (ASTM D 1329-60), DIN 53 513, and DIN 53 520) low acrylonitrile NBR and large amounts of efficient plasticisers should be used.

For optimum low-temperature behaviour (according to the Brittleness Test (ASTM D 746-64 T) and Cold Bending Test (ASTM D 736-54T)) low acrylonitrile NBR active carbon blacks (no plasticiser) are preferred. However, where inactive fillers are used, large amounts of efficient plasticisers should be employed.

6.5. Compression Set

The relations between compression set and compound formulation have been considered in detail by Jahn and Bertram.[112] In particular they considered the implications of measuring the recovery at the test temperature and the differences in respect of reversible and irreversible relaxation processes as functions of temperature. In practice long-term compression set resistance at temperatures in the region of 120 °C or even 150 °C is being specified more and more often. Apart from peroxide cures, cures with combinations of thiuram disulphide with sulphur donors (N,N'-dithio-bis-morpholine or caprolactam disulphide) have given the best results where long-term compression set is concerned.

6.6. Swelling Resistance

The difficulties facing the compounder are greater here than anywhere else. In the majority of cases the vulcanisates must have definite degrees of resistance to several oils and liquid fuels and must, at the same time, meet specific low-temperature flexibility and mechanical property requirements. The most important factors are the choice of the NBR grade (which may possibly be blended with other rubbers), the type and amount of filler, and the type and quantity of plasticiser. The interrelations are so complex that they cannot be considered at length in an article which represents merely a survey. Well-known NBR manufacturers supply computer studies of this subject in which the main factors have been evaluated mathematically, with the result that the number of experiments which have to be performed is greatly reduced.

6.7. Ozone Resistance
NBR vulcanisates are only moderately resistant to ozone. The ozone resistance can be increased effectively by blending the NBR with PVC or EPDM. If, for technical reasons, this is not possible, all that can be done is to add large amounts of p-phenylene diamines or the non-staining antiozonant AFD (which is based on enol ether) together with waxes. No explanation has yet been found for the fact that antiozonants have so little effect in NBR vulcanisates. Grades with low acrylonitrile contents can be given more protection than those containing large amounts of acrylonitrile.[113]

7. PRESENT PROBLEMS AND FUTURE PROSPECTS

7.1. Heat Resistance
The opportunities for improving the heat resistance of NBR vulcanisates through the use of improved polymers and optimised compounds have been described in Sections 4.6 and 6.3. The progress in this field has been sufficient to ensure that NBR will retain many automotive and general engineering applications in which improved heat resistance is necessary.

The developments of recent years, which have given a new lease of life to an inexpensive and proven polymer, are investigated from the standpoint of motor vehicle manufacturers by Walter.[114] Whether further substantial progress will be possible is uncertain. Improvements could result from the use of hydrogenated NBR (cf. Section 4.7), but it is not certain that this material will be produced industrially. The extent to which demands will become more exacting in the future is another unknown quantity. At the moment little can be said by the motor vehicle industry, as the design of motor vehicles is expected to undergo considerable changes in consequence of the world energy shortage and present and future national legislation (relating to such matters as speed and fuel consumption limits, exhaust pollution, and the composition of fuels).

7.2. Future Fuels
NBR has gained many of its applications through the ability of its vulcanisates to withstand liquid fuels. Future developments in fuels may therefore be of considerable importance.

7.2.1. Effect of Oil Scarcity
Such developments are likely to arise not just from changes in technology, but also from legislation aimed at protecting the environment, from energy shortages, and possibly from the policies of individual countries.

Although many different forecasts can be made, it is certain that oil will continue to become scarcer, with the result that its price will increase drastically and alternative fuels will become more important.[115] Legislation in a number of countries has reduced the tetraethyl lead content of petrol. Aromatics are therefore being added to compensate for the reductions in octane numbers which would otherwise result. Low-lead petrol in the Federal Republic of Germany has an aromatics content of 40–45 %, whereas the corresponding figure in the United States is 33–36 %.

Compared with American fuels, European fuels therefore exert a more pronounced swelling action on rubber goods. For reasons of industrial hygiene it is unlikely that further increases in the aromatic contents of petrol will be forthcoming.

7.2.2. Alcohol Addition
It is probable that methanol or ethanol will be used to some extent as a fuel substitute in order to raise octane numbers and save conventional fuels. A number of technical problems (arising from phase separation of fuels, increased vapour pressure values, and cold start difficulties) remain to be solved in this connection.

In a hydrocarbon/methanol mixture the proportion of relatively stable hydrogen bonds is reduced. In addition, polar dipole–dipole interactions take place between the alcohol and the nitrile group of the rubber, with the result that NBR vulcanisates swell much more severely in hydrocarbon/alcohol mixtures than they do in the individual fluids.[116,117] Apart from this, all the diffusion processes are accelerated to the extent that the polymer network is loosened. Accordingly the addition of alcohols to hydrocarbon fuels may be expected to increase the extent to which liquid fuels diffuse through rubber goods.

At the moment it appears that new fluoro rubbers, and possibly epichlorohydrin rubber, could offer a solution to these problems. On the other hand NBR manufacturers are working along lines which do not entail a renunciation of NBR. So many questions are still unanswered that any prediction of the future must still be hazardous.

7.2.3. The Sour Gas Problem

The so-called sour gas problem[118] is another consideration affecting the use of NBR vulcanisates in contact with liquid fuels. Overstored fuels, and those consumed by engines with fuel injection, contain hydroperoxides which exert hardening and softening actions on NBR and epichlorohydrin rubber respectively. Only fluoro rubber is not attacked. Dunn et al.[119] have found, however, that NBR goods are fully able to meet the 'sour gasoline' resistance requirements if special compounds are used—compounds containing sulphur donor crosslinking systems, silica, cadmium oxide, and special antioxidants. The compounding principles are similar to those which give the best possible resistance to ageing in hot air. The use of NBR grades containing polymer-bound antioxidants has not been found to give any advantage.

It should be added that the sour gas problem has so far arisen only in the United States and Japan and that it is evidently not experienced in connection with all fuel injection engines. Doubt therefore remains as to whether it is a universal problem.

7.3. Future Processing Methods

The possibility of the use of liquid reactive NBR grades was mentioned in Section 4.9. It was pointed out, however, that the industrial manufacture of these materials could not be taken for granted and that, in any case, it would have to be preceded by developments in processing. It is equally uncertain whether the development of thermoplastically processable NBR will be possible. The main difficulty in this connection is that many NBR applications involve service temperatures of 90–120 °C. A thermoplastically processable NBR would therefore have to have dimensional stability at fairly high temperatures. The processing of NBR by the powdered rubber process is regarded as a far less distant possibility, and also as very interesting commercially. Very considerable progress in this area has resulted from close cooperation between Lehnen[120] and the processing plant industry; thus it appears that the direct processing of NBR compounds in powder form will permit savings in total production costs of around 50%. The process has been substantially perfected for injection moulding and is now being adapted to extrusion.

7.4. Prospects

In some of its applications NBR has been replaced by other special rubbers that are superior to it in individual respects, or in combinations of several properties. Thus acrylate, silicone, and fluoro rubbers have already been used for a number of years for many types of shaft oil seal, which were formerly made with NBR. The change was dictated by higher engine temperatures and speeds and alterations to the composition of lubricating oils, together with longer guarantee periods and intervals between oil changes. Even the use of NBR for certain rubber parts which come into contact with liquid fuels has been replaced by that of epichlorohydrin polymers, which offer more favourable combinations of resistance to swelling, low-temperature flexibility, and resistance to ozone. It is here, above all, that the future of NBR will depend very much on the fuels of the future. On the whole, however, the position of NBR as the standard polymer for rubber goods that are exposed to oils, liquid fuels, and greases is unlikely to be seriously challenged. In those applications where the goods are only just equal to the demands made on them, NBR manufacturers and the rubber industry are endeavouring to ensure the continued use of this material by developing new grades and processing methods and working out new and more suitable recipes.

ACKNOWLEDGEMENT

The author wishes to thank Bayer AG for permission to publish this paper.

REFERENCES

1. HOFMANN, W., *Rubber Chem. Technol.*, **36**, 1963, 1.
2. DUNN, J. R., COULTHARD, D. C. and PFISTERER, H. A., *Rubber Chem. Technol.*, **51**, 1978, 389.
3. SHEN, M. C. and EISENBERG, A., *Rubber Chem. Technol.*, **43**, 1970, 95.
4. BERTRAM, H. H. and BRANDT, D., *Rubber Chem. Technol.*, **45**, 1972, 1224.
5. International Institute of Synthetic Rubber Producers, *The Synthetic Rubber Manual, 8th edition*, 1980.

82 H. H. BERTRAM

6. NORDSIEK, K. H., *International Rubber Conference, Munich*, September 1978.
7. YELLAND, P., *Plastics and Rubber International*, **2**, 1977, 67.
8. EVANS, C. W., *IRI Second Annual National Conference, Blackpool*, 1974.
9. KONRAD, E., *Gummi-Ztg.*, Jubilee No. 13, 1936.
10. BRYANT, C. L., *J. Inst. Rubber Ind.*, **4**, 1970, 202.
11. BROWN, H. P., *Rubber Chem. Technol.*, **36**, 1963, 931.
12. HALLENBECK, V. L., *Rubber Chem. Technol.*, **46**, 1973, 78.
13. *Polysar Progress*, Polysar Ltd, Canada, March–April 1977, p. 1.
13a. JONES, R. H., WALKER, J. and WEIR, R. J., *4th Australian Rubber Technol. Convention*, 1977.
14. KLINE, R. H. and MILLER, J. P., *Rubber Chem. Technol.*, **46**, 1973, 96.
15. MEYER, G. E., KAVCHOK, R. W. and NAPLES, F. J., *Rubber Chem. Technol.*, **46**, 1973, 106.
16. HORVATH, J. W., *SAE Automotive Engng. Meeting, Detroit*, October 1975, paper No. 750960.
16a. HORVATH, J. W., GRIMM, D. C. and STEVICK, J. A., *J. Elastomers and Plastics*, **7**, 1975, 337.
17. HORVATH, J. W., PURDON, J. R., MEYER, G. E. and NAPLES, F. J., *Applied Polymer Symposium*, No. 25, 1974, 187.
18. HORVATH, J. W. and BUSH, J. L., SAE paper No. 770863, 1978.
19. SCOTT, G., *Europ. Polym. J. Suppl.*, 1969, 189.
20. SCOTT, G., *Macromol. Chem.*, **8**, 1973, 319.
21. SCOTT, G. and SMITH, K. V., *Rubber Chem. Technol.*, **52**, 1979, 949.
22. SCOTT, G., *International Rubber Conference, Brighton*, May 1977.
23. WEINSTEIN, A. H., *Rubber Chem. Technol.*, **50**, 1977, 641.
24. WEINSTEIN, A. H., *Rubber Chem. Technol.*, **50**, 1977, 650.
25. Firestone, US Patent 3 882 094, 1975.
26. Firestone, US Patent 3 993 855, 1976.
27. Bridgestone Tire and Rubber Co., Japanese Patent 67/47 897, 1967.
28. Nippon Zeon, Japanese Patent 78/39 744, 1978.
29. Bayer AG, DT 2 539 132, 1977.
30. EDWARDS, D. C. and SATO, K., *Rubber Chem. Technol.*, **52**, 1979, 84.
31. WAGNER, M. P., *Rubber World*, **164**(5), 1971, 46.
32. DANNENBERG, E. M. and COTTEN, G. R., *International Symposium on Elastomer Reinforcement, Le Bichenberg*, September 1973.
33. Union Carbide Corporation (UCC), Bulletin TT-TL-1438, June 1976.
34. WOLFF, S., *Rubber Chem. Technol.*, **50**, 1977, 447.
35. DRAKE, R. S. and McCARTY, W. J., *Rubber World*, **159**(1), 1968, 51.
36. HUMPIDGE, R. T., MATTHEWS, D., MORELL, S. H. and PYNE, J. R., *Rubber Chem. Technol.*, **46**, 1973, 148.
37. KALFOGOLOU, N. K. and WILLIAMS, H. L., *J. Appl. Polymer Sci.*, **17**, 1973, 1377.
38. SIEBERT, A. R., *J. Elastomers and Plastics*, **8**, 1976, 177.
39. ROWE, E. H., SIEBERT, A. R. and DRAKE, R. S., *Mod. Plast.* **47**(8), 1970, 110.
40. BOULONNAIS, D., *Rev. Gen. Caoutch. Plast.*, **51**, 1974, 341.
41. FURUKAWA, J. and ISEDA, Y., *J. Polym. Sci. Polymer Letters*, **7**, 1969, 47.
42. FURUKAWA, J., *et al.*, *J. Polym. Sci. Polymer Letters*, **7**, 1969, 561.

43. TAKAMATSU, T., ONISHI, A., NISHIKADA, T. and FURUKAWA, J., *Rubber Age*, **105**(6), 1973, 23.
44. MUKHERJEE, D. P. and GOLDSTEIN, C., *Rubber Chem. Technol.*, **46**, 1973, 1264.
45. FURUKAWA, J., NISHIOKA, A. and KOTANI, T., *Polymer Sci.*, B 8/1970, 24.
46. LASIS, E., BUCKLER, E. J. and DUNN, J. R., (to Polysar Ltd), Canadian Patent 1 014 295, 1977.
47. LASIS, E. and BUCKLER, E. J. (to Polysar Ltd) Canadian Patent 185 313, 1973.
48. LASIS, E. and BUCKLER, E. J. (to Polysar Ltd), Canadian Patent 185 359, 1973.
49. LASIS, E. and BUCKLER, E. J. (to Polysar Ltd), Canadian Patent 185 313, 1973.
50. LASIS, E. and BUCKLER, E. J. (to Polysar Ltd), DT 2 613 050, 1976.
51. DEVIRTS, E. A., IZMAILOVA, L. V. and MOISEEV, V. V., *Soviet Rubber Technol.*, **31**(12), 1972, 13.
52. BERTRAM, H. H., unpublished work, Bayer AG.
53. Polysar International S.A., *Plast. Rubb. News*, March 1979, 41.
54. SHAPIRO, Y. E., SHVETSOV, O. K. and USTAVSHCHIKOV, B. F., *Vys. Soed. B.*, **18**, 1976, 736.
55. GATTI, G. and CARBONARO, A., *Makromol. Chem.*, **175**, 1974, 1627.
56. KOMA, Y., IIMURA, K. and TAKEDA, M., *J. Polym. Sci.*, *Polym. Chem. Ed.*, **10**, 1972, 2983.
57. *Ullmann's Enzyklopädie der technischen Chemie*, Vol. 13, Verlag Chemie, Weinheim, 1977, p. 611.
58. DINGES, K., *Kautschuk u. Gummi. Kunststoffe*, **32**, 1979, 748.
59. CORISH, P. J. and POWELL, B. D. W., *Rubber Chem. Technol.*, **47**, 1974, 481.
60. DINGES, K., *Methodicum Chimicum, Polymere Werkstoffe*, Verlag Thieme, Stuttgart, to be published.
61. STOLLFUSS, B., Technical Information Bulletin, Perbunan N, No. 2.7.1., September 1977, Bayer AG.
62. MITCHELL, J. M., *Gummi. Asbest. Kunststoffe*, **30**, 1977, 498.
63. RADKE, R. and BOETTCHER, E., *Gummi. Asbest. Kunststoffe*, **30**, 1977, 72.
64. WHITTINGTON, W. H., *J. IRI*, **9**(4), 1975, 151.
65. MITCHELL, J. M., *J. Elastomers and Plastics*, **9**, 1977, 329.
66. WOODS, M. E. and DAVIDSON, J. A., *Rubber Chem. Technol.*, **49**, 1976, 112.
67. MASTROMATTEO, R. P., MITCHELL, J. M. and BRETT, T. J., *Rubber Chem. Technol.*, **44**, 1971, 1065.
68. STOLLFUSS, B. Technical Information Bulletin, Perbunan N, No. 1.2.2., March 1977, Bayer AG.
69. ABRAMS, W. J., *Rubber Age*, **91**(2), 1962, 255.
70. SCHWARZ, H. F. and EDWARDS, W. S., *Appl. Polym. Symp.*, **25**, 1974, 243.
71. BHOWMICK, A. K. and DE, S. K., *Rubber Chem. Technol.*, **53**, 1980, 107.
72. ZAKRZEWSKI, G. A., *Polymer*, **14**, 1973, 347.
73. KRONMAN, A. G. and KARGIN, V. A., *Polymer Sci. USSR*, **8**, 1966, 1878.
74. MANN, J. and WILLIAMSON, R., *Physics of Glassy Polymers*, J. Wiley and Sons, New York, 1973.
75. KÜHNE, G. F., *SPE Techn.*, paper 17, 1971, 491.
76. MATSUO, M., NOZAKI, C. and JYO, Y., *Polymer Eng. Sci.*, **9**, 1969, 197.
77. JORDAN, E. F., ARTYMSHYN, B. and RISER, G. R., *J. Appl. Polym. Sci.*, **21**, 1976, pp. 2715, 2737 and 2757.

78. BERTRAM, H. H., 1963, unpublished.
79. KHANIN, S. E., ANGERT, L. G., KULEZNEV, V. N. and SHASHKOV, S., *Kauch. i. Rezina*, **33**(1), 1974, 30.
80. JORGENSEN, A. H. and FRAZER, D. G., *Appl. Polymer Symp.*, **7**, 1968, 83.
81. ZATEEV, V. S., KUZMINSKII, A. S. and FRENKEL, R., *Sh. Proizvad Shin Rezinotekh Izedii Ref. Sb.*, 1972, 4; *C.A.* **78**, 1973, 137 606 g.
82. LANDI, V. R., *ACS Macromol. Secr. Symposium, Chicago*, August 1973.
83. LANDI, V. R., *Rubber Chem. Technol.*, **45**, 1972, 222.
84. SCHOLTAN, W., LANGE, H., CASPER, R., POHL, W., WENDISCH, D. and MAYER-MADER, R., *Angewandte Makrom. Chemie*, **27**, 1972, 1.
85. NAKAJIMA, N. and HARRELL, E. R. *Rubber Chem. Technol.*, **53**, 1980, 14.
86. TOLSTUKHINA, F. S. and KOLESNIKOVA, N. N., *Rev. Gen. Caoutch. Plast.*, **48**, 1971, 1239.
87. STOLLFUSS, B., unpublished.
88. GILLER, A., *International Rubber Conference, Kiev*, 1978, paper C 17.
89. KOSFELD, R. and BOROWITZ, J., *5 Conférence Européenne des Plastique et des Caoutchoucs, Paris*, 1978, Vol. 2.
90. WIJAYARTHNA, B., CHANG, W. V. and SALOVEY, R., *Rubber Chem. Technol.*, **51**, 1978, 1006.
91. NAKAJIMA, N. and COLLINS, E. A., *Rubber Chem. Technol.*, **49**, 1976, 52.
92. NAKAJIMA, N. and COLLINS, E. A., *Rubber Chem. Technol.*, **48**, 1975, 615.
93. BITTEL, P. A., *Elastomerics*, **112**(4), 1980, 44.
94. GOODRICH, B. F., *Manual HM-9*.
95. BERTRAM, H. H. and STOLLFUSS, B., Technical Information Bulletin, Perbunan N, No. 2.2.1., November 1977, Bayer AG.
96. LEIBBRANDT, F., Technical Information Bulletin, Perbunan N, No. 2.2.2., August 1978, Bayer AG.
97. DT 2138582 (Uniroyal), 1972.
98. DT 2223808 (Uniroyal), 1972.
99. DT 2902412 (Du Pont), 1979.
100. LEE, T. C. P. and MORELL, S. H., *RAPRA Research Report*, 201, 1972; *Rubber Chem. Technol.*, **46**, 1973, 483.
101. KOSINSKA, K. Z. and LEE, T. C. P., *RAPRA Members J.*, **1**(6), 1973, 146.
102. BERTRAM, H. H., Technical Information Bulletin, No. 16, 1967, Bayer AG.
103. BERTRAM, H. H., Technical Information Bulletin, No. 17, 1967, Bayer AG.
104. STOLLFUSS, B., Technical Information Bulletin, Perbunan N, No. 2.8.1.1., December 1977, Bayer AG.
105. PFISTERER, H. A. and DUNN, J. R., *J. Elastomers and Plastics*, **7**, 1975, 427.
106. DUNN, J. R. and PFISTERER, H. A., *J. Elastomers and Plastics*, **9**, 1977, 193.
107. COULTHARD, D. C. and GUNTER, W. D., *J. Elastomers and Plastics*, **9**, 1977, 131.
108. BLOW, C. M., *Rubber J.*, **155**(7), 1973, 18.
109. PAULIN, D. A., *Rubber Age*, **101**(10), 1969, 69.
110. ABDELKADER, M. H. and YOUSIF, S. M., *Plaste u. Kautschuk*, **21**, 1974, 521.
111. ENGELMANN, E., *Kautschuk u, Gummi. Kunststoffe*, **25**, 1972, 538.
112. JAHN, H. J. and BERTRAM, H. H., *102nd ACS meeting, Cincinnati*, 1972.
113. STOLLFUSS, B., Technical Information Bulletin, Perbunan N®, No. 1.2.2., March 1977, Bayer AG.

114. WALTER, G., *Gummi. Asbest. Kunststoffe*, **28**, 1975, 126 and *Rubber Chem. Technol.*, **49**, 1976, 775.
115. FRIBERG, G., *Scandinavian Rubber Conference, Copenhagen*, 1979.
116. PIAZZA, S., SANTARELLI, G. and PASSARINI, N., *Scandinavian Rubber Conference, Copenhagen*, 1979.
117. HERTZ, D. L., *Rubber and Plastics News*, 4 February 1980, 22.
118. NERSASIAN, A., *Rubber and Plastics News*, 26 June 1978, 14.
119. DUNN, J. R., PFISTERER, H. A. and RIDLAND, J. J., *Gummi. Asbest. Kunststoffe*, **33**, 1980, 296.
120. LEHNEN, J., *Conf. Plastics and Rubber Institute, Southampton*, April 12, 1978.

Chapter 4

ETHYLENE–PROPYLENE RUBBERS

L. CORBELLI

Montedison S.p.A., Ferrara, Italy

SUMMARY

Ethylene–propylene elastomers have become and will continue to be widely used products of superior quality. After a slow start, their position on the market is now clear and it is unlikely they will be ousted from this position.

After giving some information on the production cycle and the intrinsic properties of these elastomers, there follows a classification of the types which are available. These cover practically all the requirements of the industry, and bearing in mind the need for ever greater economies in terms of energy and work, only the most suitable forms have been put on the market.

For the technologist, knowledge has been updated on curing systems, compounding and the more interesting stages of processing, giving particular attention to mixing.

The applications mentioned are the more well known but newer ones in the automotive sector, building, electrical insulation, domestic appliances, agriculture, etc. are also mentioned. Information is also given on uses of ethylene–propylene elastomers where the products do not undergo the traditional processing of the rubber industry, but are used as additives to various materials such as plastics, lubricating oils, bitumen and waxes, in order to modify or improve certain properties.

1. INTRODUCTION

The first amorphous ethylene–propylene copolymers with elastomeric properties were developed and laboratory tested toward the middle of the 1950s by Natta *et al.*[1-3]

The chemically saturated nature of these copolymers immediately made researchers face the problem of crosslinking by methods that would be accepted by the traditionally conservative rubber industry. The use of organic peroxides soon became the most promising solution, notwithstanding the cost of considerable research work necessary to improve the efficiency of the crosslinking process by certain substances acting as 'coagents'.

In the meantime terpolymers were being developed. Because they contained a small amount of unsaturated monomer, crosslinking was possible with peroxides and also with traditional accelerator and sulphur based systems. The advent of the more versatile, and consequently the more accepted terpolymers, did not however, cause the disappearance of the copolymers: this was because the aversion to peroxide cure was decreasing, and the two types could therefore coexist on the market.

In fact, applications now exist where copolymers and their compounds are to be preferred to terpolymers for both technical and cost reasons, as explained further on in this chapter.

Ethylene–propylene copolymers and terpolymers can certainly be considered the most interesting family of elastomers to appear in recent years and they are also well accepted by the rubber industry.

Their exceptional resistance to atmospheric agents in general (particularly to ozone) and to many chemicals, their excellent low- and high-temperature behaviour, excellent dielectric characteristics and the competitive cost of their compounds make these rubbers suitable for quality applications where these properties are required.

Ethylene–propylene elastomers are commonly identified by the abbreviation EPM for copolymers and EPDM for terpolymers. The letter M, in accordance with ASTM nomenclature, indicates the class of elastomers with a methylene type saturated chain. The polymers will be described henceforth by these abbreviations.

2. PRODUCTION

2.1. Monomers

The base monomers used, as the name suggests, are ethylene and propylene. They are manufactured directly from petroleum cracking, are available in large quantities and are amongst the least expensive monomers. The quantity ratio of the two monomers confers particular properties on the final elastomer, and it is therefore very important that it should be clearly defined and constant during the polymerisation stage. In practice this requires accurate control, particularly because the reactivity of the two monomers is very different.

2.2. Third Monomer

As already mentioned, a third, diene-type monomer is copolymerised in small quantities with the ethylene and the propylene so as to introduce a few unsaturated points which will allow a sulphur/accelerator cure. The choice of third monomer was not an easy task because the following requirements had to be met:

1. A maximum of two double bonds, one of which had to be polymerisable, the other curable at a later stage
2. Reactivity similar to that of the two base monomers
3. A random polymerisation in the principal chain for uniform distribution
4. Molecular weight not too high—so as to avoid problems during separation of the non-reacted monomer during the final stages
5. The final polymer to have an acceptable cure rate

Of the many monomers tested, those which are currently used in commercial production are ethylidene norbornene (ENB), 1,4-hexadiene (1,4-HD) and dicyclopentadiene (DCPD). Their formulae are given in Fig. 1.

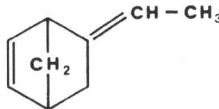

5 – ETHYLIDENE – 2 – NORBORNENE

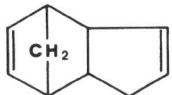

DICYCLOPENTADIENE

$$CH_2 = CH - CH_2 - CH = CH - CH_3$$
1,4 – HEXADIENE

FIG. 1.   Third monomers for EPDMs.

TABLE 1
CATALYST AND COCATALYST FOR ETHYLENE–
PROPYLENE COPOLYMERISATION

Catalyst	Cocatalyst
$TiCl_2$	$Al(C_6H_{13})_3$
$TiCl_3$	$Al(C_2H_5)_2Cl$
$TiCl_4$	$Al(i\text{-}C_4H_9)_2Cl$
$VOCl_3$	$Al(i\text{-}C_4H_9)_3$
VCl_4	$Al_2(C_2H_5)_3Cl_3$
VCl_3	$Al(n\text{-}C_6H_{13})_3$
VAc_3	$HCl_2AlO(C_2H_5)_2$
$VOAc_2$	
$VOAc_2Cl$	
$VOAcCl_2$	
$VO(O\text{-}n\text{-}C_4H_9)_3$	
$VO(OC_2H_5)_2Cl$	
$VO(OC_2H_5)Cl_2$	
$VO(OC_2H_5)_3$	

Ac = Acetylacetonate

The most commonly used monomer at the moment is ENB because it is the one which best meets the above requirements. Compared to ENB, 1,4-HD gives the polymer a slightly slower cure rate. Amongst the advantages of DCPD are low cost and a good copolymerisation rate, but its polymers are slow in crosslinking and are very branched.

2.3. Catalyst Systems

The catalysts are those developed by Natta *et al.* They consist of compounds of titanium and vanadium; the cocatalysts are alumino-alkyls (Table 1).

Titanium catalysts were soon abandoned as they produced elastomers with poor elasticity. However, they have recently been reconsidered and elasticity values have been obtained which are comparable with the best elastomers obtained with vanadium compounds.[4] Coupling a vanadium compound with an aluminium compound obviously opens the way to many combinations and past research has been pointed towards the most suitable one. It is in fact on catalyst choice that process efficiency and quality of the finished product depend.

The molecular weight regulator is also part of the catalyst system. Of the many substances tested, those which have been found to be most suitable for industrial use are hydrogen and zinc alkyls.[5,6]

Much work has been done in the past to discover suitable substances which, by reactivating the catalyst, would increase efficiency with considerable technical and cost advantages.

2.4. Processes

Currently adopted processes are basically of two types: solution and suspension. In the former, the reaction medium is a solvent (e.g. hexane) to which ethylene, propylene and possibly a third monomer are added. These materials must be suitably treated as the presence of water or other polar substances would destroy the catalysts.

As soon as the catalysts are added, reaction begins and the polymer is produced and immediately dissolves in the solvent. The solution is carefully washed to remove any residual catalyst and then the polymer is separated from the solvent by boiling water. It remains in suspension in the water in the form of crumbs. The polymer is finally separated from the water, dried and baled.

In the suspension process, the reaction medium consists of liquid propylene to which are added ethylene, the third monomer if necessary, and the catalysts; the polymer thus formed is insoluble in propylene and remains suspended therein.[7]

Because of the high concentration of monomer in this system greater specific production is obtained, compared to the solution process, together with higher monomer conversion. There are no complications due to viscosity because of the fact that the polymer is insoluble.

One basic characteristic of the suspension process is that the reaction heat may be removed by evaporation of the monomers which are subsequently compressed, condensed and recycled to the reactor. It is therefore possible to guarantee temperature control throughout the whole of the reactor. In the solution process, removal of reaction heat generally occurs through wall cooling thus leading to difficulties in accurate temperature control inside the reactor itself.

The suspension process will normally produce in the range of 20–60 $\times 10^3$ g polymer per gramme of vanadium catalyst (depending on the type of elastomer). The slurry is discharged continuously from the reactor and it is not necessary to remove the catalyst residues because of the small quantities involved. The non-converted monomers are steam separated, recovered, purified and recycled to the reactor. The polymer, which is now in an aqueous suspension, is separated from the water by a shaker screen, and is then dried and baled.

In the solution process the concentration of the polymer in the solvent is normally 5–6% in weight. Because of the high viscosity of the solution greater concentrations would lead to difficulties in agitation, thermal exchange and diffusion of the monomers and adequate control of the reaction would be impossible. In the suspension process the concentration of polymer in the reactor could easily reach 5–6 times the maximum solution quantities, with undoubted technical and cost advantages.

With the solution process, the joint effects of high solution viscosity and the dissolving action of the solvent on the concentration of the reagents lead to a considerable decrease in yield from the catalyst. The yield is therefore many times lower than the yield obtainable from the suspension process.

The suspension process will produce polymers with a very high molecular weight together with an extremely high polymerisation yield. The solution process becomes more and more critical and uncontrollable as the molecular weight of the polymer increases. Consequently, higher molecular weight polymers cannot be manufactured by this process, both for technical and for cost reasons.

Thanks to the excellent reaction control, the suspension process will produce polymers with a consistent ratio of ethylene to propylene, which in turn means consistency in physical properties. Even the alternation between the monomers is optimum.

These comments also apply to terpolymer systems. The third monomer is incorporated into the growing molecular chain in a well controlled and constant ratio, its distribution in the polymeric chain is very uniform thus giving greater efficiency during curing.

With the suspension process, it is also possible to control the molecular weight distribution. Because of the absence of smaller fractions this allows highly efficient peroxide crosslinking in EPMs and better usage of the third monomer in EPDMs. Improved control of molecular weight distribution has positive effects also on the rheological properties of the polymers. In fact, minimum variability of this characteristic considerably reduces some of the drawbacks common to all elastomers, e.g. variation in extrusion rate, swelling and shrinkage.

3.   STRUCTURE AND PROPERTIES OF THE POLYMERS

3.1. Composition
The composition of EPMs is defined by the percentage content by weight of propylene. This is derived from the most commonly used analysis method,

infra-red spectroscopy, which determines the methyl group concentration of the bonded propylene: the same method is used for EPDMs.

All the types of EPM and EPDM currently available on the market have a propylene content of between 25 and 55% by weight. Usage of a low propylene content will produce polymers with high green strength; the medium and higher values of propylene will produce softer and more elastic polymers.

Infra-red spectrometry will also determine the quantity of the bonded third monomer; this is between 2 and 5% by weight for most EPDMs. In recent years, however, high crosslinking rate types have been developed in which the quantity of third monomer can reach 10%.

3.2. Molecular Weight

In line with all commercial elastomers, molecular weight is directly and conventionally correlated with the Mooney viscosity. Unfortunately, the various manufacturers of EPMs and EPDMs adopt different measurement conditions and comparison cannot always be immediate. It is good practice to limit the measurement at 100 °C to those polymers which have viscosities no higher than 60–70 points ML (1 + 4); high viscosity polymers should be measured at 121 °C or, better still, at 125 °C. The time (1 + 4) minutes gives sufficiently reproducible results; the greater reproducibility obtained at (1 + 8) minutes is perhaps not justified for practical purposes due to the longer test time. Viscosities of commercial polymers range between 25 and 60 at 100 °C and do not exceed 100 at 125 °C. The higher molecular weight polymers are sold with the addition of a certain quantity of oil.

Under fixed conditions during polymerisation, the molecular weight of the polymer increases as the ethylene content increases.

3.3. Molecular Weight Distribution

In recent years, it has been possible to demonstrate the correlation between the molecular weight distribution (MWD) of a polymer and its rheological behaviour. It is obviously necessary, therefore, to keep this parameter under control during the polymerisation process. By its very nature, the catalyst system used to manufacture ethylene–propylene elastomers will produce polymers with a fairly narrow MWD which can, however, be varied between certain limits. In commercial polymers, the M_w/M_n ratio, which is a good indication of the MWD, generally varies between 3 and 5.

Some EPMs and EPDMs available on the market have a much wider distribution but they in fact consist of a mechanical blend of two or more polymers with different molecular weights.

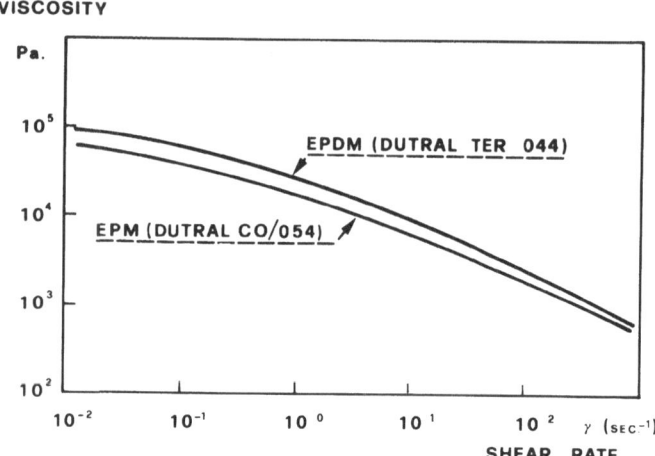

FIG. 2. Flow curves at 150 °C of ethylene–propylene elastomers.

3.4. Rheological Properties

During the various stages of processing a wide range of temperatures and shear rates may be encountered. Good characterisation from this point of view can be achieved by determining the flow curve of a particular compound with a capillary rheometer (Fig. 2). This method gives useful information but the ultimate verification must be carried out on the processor's machines.

Small amounts of branching can exist in ethylene–propylene elastomers, particularly in EPDMs, and to an extent this depends on the type of third monomer used. It is logical to expect that the branching should influence the rheological behaviour of the polymer, but unfortunately there is a lack of practical methods which can be used to measure the effects of branching. A more obvious and important characteristic of ethylene–propylene elastomers is what is known as green strength. The term 'green strength' indicates the tensile strength of the uncured compound and may be measured by putting normal dumb-bells under traction in the equipment used for tensile testing of cured elastomers. It has already been mentioned that the content of propylene in EPMs and EPDMs can vary from around 25 to 55 %. In this range the resulting green strength is in the order of that shown in Fig. 3.

From the structural point of view, the level of green strength is closely correlated with the presence of ethylene blocks within the molecule; the longer the blocks the higher the green strength. These blocks can in fact

GREEN STRENGTH

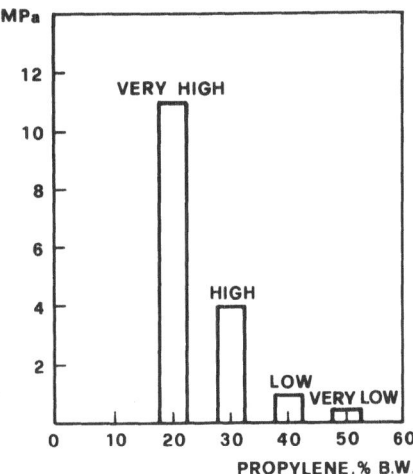

FIG. 3. Green strength of EPM and EPDM as a function of propylene content.

cause crystallisation phenomena to occur during tensile testing thus leading to a consequent increase in toughness. Other factors which influence the level of green strength are the MWD, molecular weight, and any inhomogeneous composition.[8] At room temperature, high green strength EPMs and EPDMs are very hard but at normal processing temperatures they soften and behave in the same manner as normal elastomers. A high green strength value is of assistance in handling the compounds during processing (see Section 7).

3.5. High-temperature Behaviour

This is excellent and is one of the characteristics which is a feature of ethylene–propylene elastomers. Using the finished articles at 120 °C is common practice while in special cases they can be used at temperatures up to 150 °C. EPMs are generally preferred for the more severe service requirements and such components should be peroxide cured with suitable additions of heat ageing inhibitors. Peroxide cured EPDMs also have a level of high-temperature resistance which is very similar to that of EPMs, whereas sulphur/accelerator cured EPDMs will produce vulcanisates which are less stable at high temperatures.

Referring to the composition in general, the rule is that polymers containing more ethylene are more stable at high temperatures. However, perhaps because of the basic good high-temperature behaviour of ethylene–propylene elastomers, research work to improve this important property is as yet somewhat limited.

3.6. Low-temperature Behaviour

The fact that the glass transition temperature (T_g) of ethylene–propylene elastomers depends on the composition was noted right from the first evaluations. Rebound curves as a function of temperature and composition

REBOUND

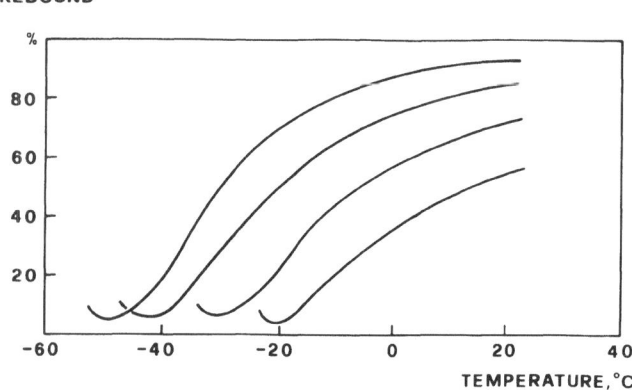

FIG. 4.   Rebound of EPM with varying propylene content (propylene content increases from left to right).

are shown in Fig. 4.[9] Since those first evaluations, work carried out in greater depth has demonstrated that there is a clear correlation between T_g and the propylene content, as shown in Fig. 5.[10] The presence of normally used quantities of a third monomer will make no substantial alteration to this behaviour.

From a more practical point of view, the TR 50 test is well known. This test determines the temperature at which an elongated test-piece frozen at a very low temperature and then unclamped, will recover 50 % of the deformation. In Table 2, such temperatures are shown for terpolymer compounds having different propylene content. Consequently, when the best low-temperature behaviour is required of an article, the propylene content of the polymer must be of the order of 40–50 % by weight.

Closely connected with low-temperature behaviour are the elastic properties of the cured polymers. Vulcanisates with good low-temperature properties show high elasticity at room temperature.

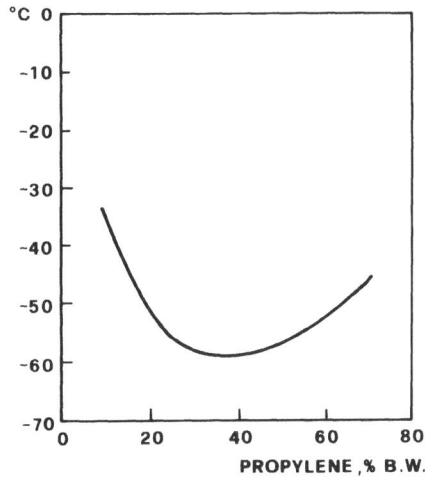

FIG. 5. EPM glass transition temperature versus propylene content.

3.7. Weathering

The resistance to weathering of ethylene–propylene rubbers is exceptional and well known. In reality this is valid mainly for black articles, even those containing very small quantities of carbon black. When the articles involved are non-black, it is necessary to protect them against UV light which would catalyse oxidative degradation. This involves a surface 'attack' which can be lessened by using UV absorber pigments (such as rutile titanium dioxide) or by using anti-UV additives. The latter, unfortunately, are not very effective.

TABLE 2
TR 50 TEST FOR EPDMS WITH DIFFERENT PROPYLENE
CONTENT

EPDM	Propylene content, % w/w	TR 50 temperature, °C
Dutral TER 058	≈ 50	−40
Dutral TER 048	≈ 40	−30
Dutral TER 038	≈ 30	−18

TABLE 3
EPM AND EPDM RESISTANCE TO CHEMICAL
AGENTS

Chemical agent	Resistance
Ozone	Excellent
Oxygen	Excellent
Water	Excellent
Alkalis	Excellent
Acids, dilute	Excellent
Acids, concentrated	Fair–good
Aliphatic solvents	Poor–good
Aromatic solvents	Poor
Oxygenated solvents	Good
Halogenated solvents	Poor
Animal oil	Fair
Vegetable oil	Fair
Mineral oil	Poor–fair
Hydraulic fluids	Good–excellent

Raw EPDM (in the form of bales for instance) can be subject to surface crosslinking when exposed to daylight for a period of even a few days. The time is further reduced if the terpolymer is oil extended or has a high third monomer content. From this point of view, EPMs are stable; in fact in many laboratories EPMs are kept for years as reference samples for checking Mooney viscometers.

3.8. Resistance to Chemicals
Because of the saturated nature of EPMs, and the predominantly saturated nature of EPDMs, this family of elastomers has considerable resistance to many chemicals. Table 3 gives a qualitative indication.

Contact with liquids in general, even if chemically inactive, can cause swelling, thereby altering the properties of the finished products. As a general rule, EPMs and EPDMs will not swell when in contact with highly polar liquids, but swelling may occur with aliphatic, non-polar or slightly polar liquids.

Water, a highly polar liquid, is therefore not compatible with EPMs and EPDMs. However, it can cause swelling, albeit only slight, if the compounds contain low molecular weight water soluble substances. In this case, the water can enter to dissolve such substances. This must be taken into account when low water absorption is required at, for example, high temperature.

TABLE 4
EPM PHYSICAL PROPERTIES

Density, kg/m^3	860
Heat capacity, cal/g °C	0·52
Thermal conductivity, cal/cm s °C	$8·5 \times 10^{-4}$
Thermal diffusivity, cm^2/s	$9·2 \times 10^{-4}$
Thermal coefficient of linear expansion, /°C	$2·2 \times 10^{-4}$
Volume resistivity, ohm cm	$5–10 \times 10^{16}$
Dielectric strength, kV/mm	30–35
Dielectric constant at 1000 cycles	2·2–2·4

3.9. Other Properties

Other physical properties of ethylene–propylene elastomers are given in Table 4 and the low specific gravity of these materials should be noted. This low specific gravity carries advantages right through to the finished product. The thermal properties are very similar to those of traditional elastomers and the electrical properties are excellent. In particular, the resistance to partial discharge has given substantial success to these elastomers in medium and high tension cable insulation; in addition, obviously, low tension insulation is also excellent.

4. TYPES OF POLYMER AVAILABLE

Current producers of ethylene–propylene elastomers are listed in Table 5. Because no agreements exist for the standardisation of production, a very

TABLE 5
EPM AND EPDM SUPPLIERS[a]

B. F. Goodrich Chemical Co.	USA
Chemische Werke Huls	West Germany
Copolymer Rubber and Chemical Corp.	USA
E. I. Du Pont de Nemours & Co.	USA
Exxon Chemical Co.	USA
International Synthetic Rubber Co.	UK
Japan EP Rubber Co.	Japan
Mitsui Petrochemical Industries	Japan
Montedison S. p. A.	Italy
Naamloze Vennootschap DSM	Holland
Societe du Caoutchouc Butyl/Exxon	France
Sumitomo Chemical Division	Japan
Uniroyal Chemical Division of Uniroyal Inc.	USA

[a] Excluding centrally planned economy countries.

high number of types of EPMs and EPDMs (at least 150) have found their way on to the market and this has been the cause of great confusion amongst users.

4.1. Classification

As we have seen in the preceding section, the most important and most significant parameters which are a characteristic of the ethylene–propylene elastomers available on the market are:

1. Molecular weight (Mooney viscosity)
2. Propylene content
3. Type of third monomer
4. Quantity of third monomer

4.1.1. Molecular Weight

The choice of molecular weight is the basis of a subdivision into three wide categories;

(a) Low–medium viscosity: ML (1 + 3) 100 °C—in the range 25–60
(b) Medium–high viscosity: ML (1 + 4) 125 °C—in the range 60–100
(c) Very high viscosity: ML (1 + 4) 125 °C—up to 200 (nominal)

Some types of the (b) group contain a certain amount of oil to ease processing. Those of type (c) are always oil extended.

Low viscosity polymers are preferable for good quality compounds with moderate filler content. High viscosities are useful for the production of low cost compounds or for improved processability.

4.1.2. Propylene Content

The bonded propylene content basically determines the green strength of the raw polymer and here again there are three categories:

(a) High propylene (50 % approx.)—low green strength
(b) Medium propylene (40 % approx.)—medium green strength
(c) Low propylene (30 % approx.)—high and very high green strength

Low green strength polymers generally produce vulcanisates with the best elastic properties. Higher green strength polymers give improved handling and are used in the production of lower cost compounds.

4.1.3. Unsaturation Level

The third monomers currently being used, as already mentioned, are ethylidene norbornene (ENB), 1,4-hexadiene (1,4-HD) and dicyclopentadiene (DCPD).

According to their third monomer content, EPDMs can be classified into three groups:

(a) Medium unsaturation content of 3–5 % in weight
(b) High unsaturation content of 5–7 % in weight
(c) Very high unsaturation content of 8–10 % in weight

Medium unsaturation polymers are those currently used in normal compounds and applications. The higher unsaturation polymers are more suitable for blending with unsaturated rubbers or to give particularly short curing cycles.

4.2. General
The combination of all these possibilities should give a very wide range of types. In practice, however, only a limited range is produced.

All EPDMs and EPMs contain small quantities of antioxidants for protection during storage. We have already mentioned the necessity of avoiding exposure of all EPDMs to light, particularly oil extended types, so as to prevent any surface crosslinking.

EPMs and EPDMs are normally sold in the form of compact bales. In recent years, particularly to overcome dispersion problems during mixing, certain higher green strength types have been produced and sold in the form of friable bales and, very recently, in crumb form.

5. CROSSLINKING SYSTEMS

Much work has been carried out in the past to find suitable crosslinking systems for EPMs and EPDMs. Whereas for the saturated EPMs research was necessary to develop new crosslinking systems, in the case of EPDMs existing systems commonly used with other unsaturated elastomers were adapted to suit the requirements of these rubbers.

5.1. Crosslinking of EPMs
5.1.1. Organic Peroxides
EPMs were produced some years before EPDMs, and for the former much work was carried out to find an acceptable crosslinking system for the rubber industry. Apart from organic peroxides, systems were studied based on chlorosulphonation and chlorination of EPM and consequent

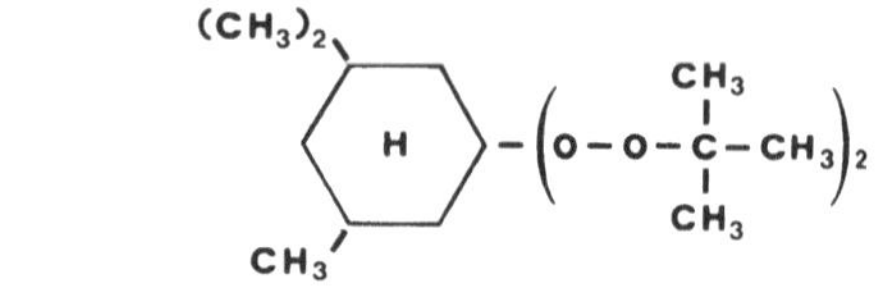

L. CORBELLI

1.1–DI–T–BUTYLPEROXY 3,5,5–TRIMETHYL
CYCLOHEXANE

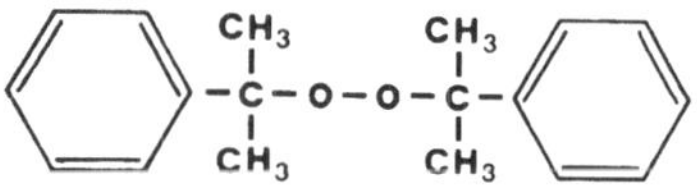

DICUMYL PEROXIDE

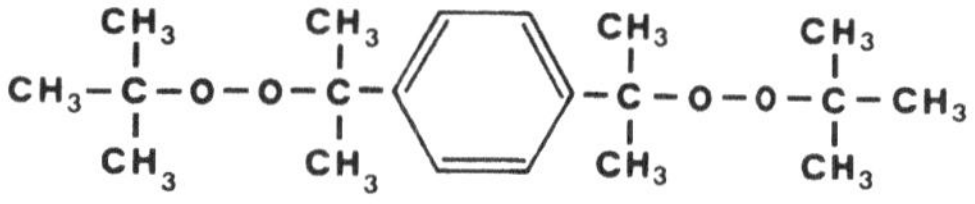

1,4 BIS (T–BUTYLPEROXY–ISOPROPYL) BENZENE

FIG. 6.   Some peroxides used as crosslinking agents.

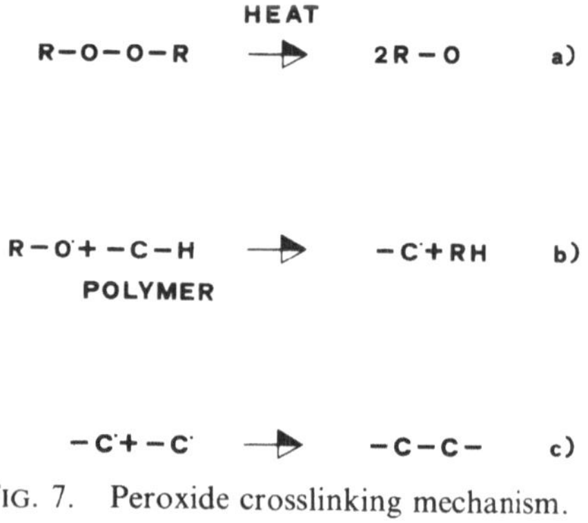

FIG. 7.   Peroxide crosslinking mechanism.

transformation of the reactive groups into double bonds suitable for crosslinking. However, the most promising system was found to be that based on organic peroxide. These substances were already known as crosslinking agents for polymers, but some difficulties existed in their commercial usage. In fact, in the 1950s di-t-butyl peroxide was already available on the market, but its high volatility made it unsuitable for crosslinking of polymers. The problem of volatility was solved later by dicumyl peroxide. However, the unpleasant odour of this restricted its use.

New peroxides were developed expressly to crosslink EPMs and each one was an improvement on the one before. A sufficient number now exists to meet most crosslinking problems (Fig. 6 lists a few peroxides for reference).

When a peroxide is heated it splits to form two radicals (Fig. 7a) each radical becomes stabilised by taking a hydrogen atom from the polymeric chain thus leaving an induced radical on a carbon atom of the same chain (Fig. 7b). The induced radical can then combine with another induced radical on another polymeric chain, thus creating the crosslink (Fig. 7c).

Unfortunately, the actual mechanism is not so simple and there are secondary reactions which sometimes considerably reduce the final degree of crosslinking.[12] Of these, the main ones are as follows:

1. Hydrolytic scission of peroxide, without creation of radicals, due to the presence of acids
2. Chain scission (depolymerisation)
3. Destruction of radicals by disproportionation
4. Grafting of fragments of peroxide on to the chain
5. A cyclisation reaction of the chain

One of the most important results of research on peroxide-curing systems is the discovery of certain substances which will improve their efficiency. The first one to have been found was sulphur[13] followed by many others called coagents, some of which are shown in Fig. 8.

In general, they consist of poly-functional substances which, during the crosslinking process, will stabilise a radical and eliminate or reduce the scission reaction described under 2 above.

When the coagent chosen is sulphur, it is possible to obtain various types of bond by increasing its quantity compared to the peroxide. The bonds may range from C—S—C to C—S_2—C to C—S_n—C and are, of course, progressively less stable at higher temperatures. The best compromise on properties is achieved with the ratio of 1 gramme-atom of sulphur to 1 mol of peroxide.[11,13] This is the optimum ratio to obtain low compression set values at high temperatures. Sulphur, however, is currently not widely used

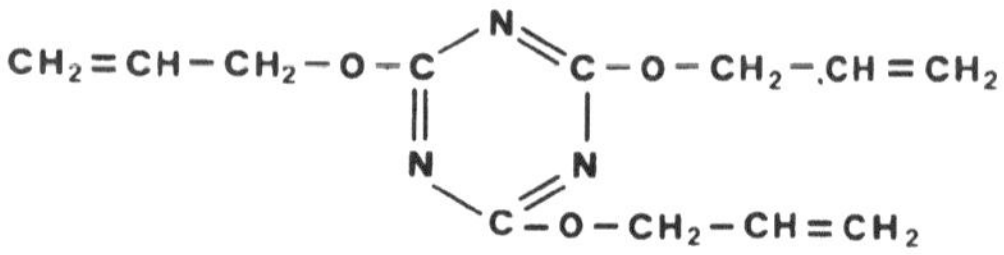

$$CH_2=CH-CH_2-O-C \overset{N}{\underset{N}{\diagdown}} C-O-CH_2-.CH=CH_2$$

TRIALLYL CYANURATE

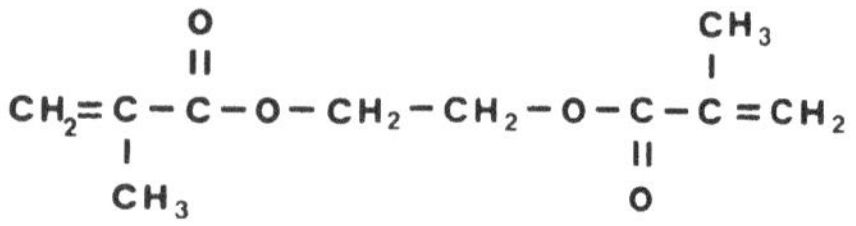

$$CH_2=\underset{CH_3}{\overset{O}{\underset{|}{C}}}-C-O-CH_2-CH_2-O-\underset{\underset{O}{||}}{\overset{CH_3}{\overset{|}{C}}}-C=CH_2$$

ETHYLENE GLYCOL DIMETHACRYLATE

$$-CH=N-N=CH-$$

DIFURFURAL ALDAZINE

FIG. 8. Some peroxide crosslinking coagents.

because of the bad odour it causes in vulcanisates and so the choice of a coagent from those given in Fig. 8 is to be preferred. The quantity compared to the amount of peroxide is less critical, but it should be regulated according to the vulcanisate properties required.

Another ingredient which is always present in the peroxide system is zinc oxide. The mechanism by which it functions is still not well understood; it probably serves to neutralise any acid substances in the system and so helps to eliminate the parasite reaction described under 1, i.e. hydrolytic scission of the peroxide. Zinc oxide is in any case indispensable when a peroxide cured article is used in a high-temperature application. Table 6 gives a few examples of EPM based formulations with peroxides. All the ingredients chosen in a formulation must be screened for their possible interference with peroxide so as to ensure that peroxide efficiency is not impaired.

Crosslinking must take place in the absence of air or oxygen as the presence of oxygen leads to degradation of the polymer and results in a sticky surface of the moulding.

Other inconveniences of peroxide crosslinking are the fixed cure conditions and greater cost compared to traditional systems. The first drawback is avoided by choosing the peroxide which is most suited to the

TABLE 6
EPM CROSSLINKED WITH DIFFERENT PEROXIDES

Dutral CO/054		100	
Zinc oxide		5	
HAF		50	
Sulphur		0·3	
Trigonox 29/40[a]	6·8		
Peroximon DC/40[b]		6·4	
Peroximon F/40[c]			4
Scorch time at 120°C			
(t_s) min	6	15–18	>20
Press curing			
Temp., °C	150	165	165
Time, min	18	30	40
Mechanical properties			
Tensile strength, MPa	17·0	19·0	18·5
Elongation at break, %	360	360	380
Modulus at 300%, MPa	13·0	15·0	14·0
Tension set at 200%, %	6	6	6
Hardness, Shore A	68	66	67
Compression set 70 h at 100°C, %	22	18	20

[a] 1,1 Di-*t*-butylperoxy-3,5,5-trimethyl cyclohexane.
[b] Dicumylperoxide.
[c] 1,4 Bis-(*t*-butylperoxy-isopropyl)-benzene.

processing and crosslinking conditions. In the case of the second, there is the compensation that the finished product will have better thermal stability and will therefore be of better quality.

Peroxide crosslinking in EPMs is more efficient with increasing levels of bonded ethylene.

5.1.2. High Energy Radiation

Radiation crosslinking, particularly with fast electrons, has come to the fore in recent years although the idea goes back many decades. It is known that EPMs are preferred for production of items which must withstand radiation, e.g. in locations such as a nuclear plant. The effect of radiation on EPMs is very small but this does not mean it is not possible to crosslink EPMs by radiation. Good practical results can in fact be obtained especially if EPMs with a high bonded ethylene content are used and if suitable 'activators' are added to the compounds. The activators are in fact the coagents used in peroxide crosslinking. In this case, as in peroxide curing, the ingredients in the compounds must be chosen with care because some can act as radiation 'absorbers'. Interest in this form of crosslinking for EPMs is however fairly limited as EPDMs are more suitable.

5.2. EPDM Crosslinking

5.2.1. Organic Peroxides

Terpolymers crosslinkable by traditional methods were originally developed to overcome some of the problems encountered in the peroxide curing of EPMs. However, compared to sulphur cured products, peroxide crosslinked EPDMs have better high-temperature resistance, lower compression set, improved electrical properties and more stable colours. They are non-staining and it is possible to reduce blooming; there is no limitation on the covulcanisation of blends of polymers of different types.

The crosslinking of EPDMs is more efficient than that of EPMs in the sense that smaller quantities of peroxide are required but in other respects what has been written for EPMs (see Section 5.1.1) applies also to EPDMs. If EPDMs are to be used at high temperatures then they must be protected by suitable antioxidants during peroxide crosslinking. The antioxidants are the same as those used with EPMs (see Section 6.7) but the use of the various coagents (mentioned in Section 5.1.1) is not so clear-cut; each must be carefully chosen for each type of EPDM and for the final application. Contrary to EPMs, EPDMs do not necessarily require the use of coagents in peroxide crosslinking. In fact, for optimum properties such as compression set, their use may even be pointless and counterproductive.

5.2.2. Sulphur and Accelerators

When EPDMs first appeared, there was difficulty in finding a good crosslinking system based on sulphur and accelerators. It was sufficient to adopt those already in use for butyl rubber as suggested by the low unsaturation level of the two families of elastomers. The commonest system was, and still is, based on mercaptobenzthiazole (0·5 phr) tetramethyl thiuram monosulphide (1·5 phr), sulphur (1·5 phr) plus zinc oxide (5 phr) and stearic acid (1 phr).

The crosslinking rates of EPDMs vary according to the type of third monomer; of those available, the fastest are those containing ethylidene norbornene, followed by 1,4-hexadiene and lastly dicyclopentadiene, as shown in Fig. 9.

In the case of EPDMs with a high third monomer content (8–10%) curing times are similar to those of the more unsaturated traditional elastomers. However, because in practice the unsaturation present in this type of EPDM is not totally used, their heat stability is inferior. As with

TABLE 7

CROSSLINKING SYSTEMS FOR EPDM

	1	2	3	4	5	6	7	8	9	10	11	12	13	14	15	16
MBT	1·5				1	1		1·5	1·5	1·5	1·5	1·3		1·2		
MBTS		0·7														
TMTD	0·8		0·8	3	0·6	0·7		0·8	0·8	1	1	0·9		0·6		
TDEDC	0·8			0·7	0·4	0·4		0·8	0·8	1	1	0·6		0·6		
DPTTS	0·8		0·8					0·8				0·7				
ZDBDC		3	2	1·5	2	2				2·5		1·8				
ZDC									0·8		1·5	0·7				
ZD$_m$C													0·8			
CBS							0·5			1			2			
DTDM			2										2			
DBTU						1							2			
Thiocarbanilide																
Vocol S																
DPG							0·3									
Deovulc EG 3															4	4·5
Vulkalent A							1									
END–75 (Masterbatch of ethylene thiourea (Na 22))								0·5	0·5		0·7	0·7				
Sulphur	1·5	2		0·5	1·5	2	2	2	2	1·7	2	1·3	2	1·5	1·5	2

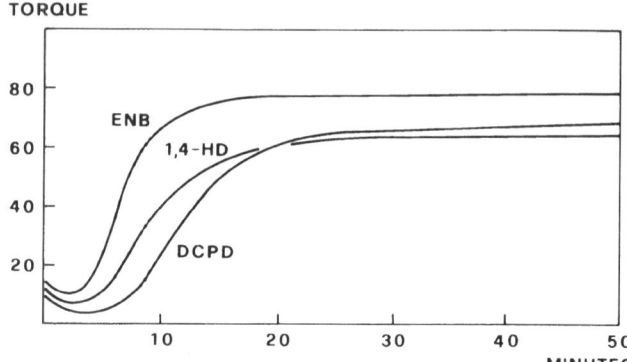

FIG. 9. Rheometer curves of EPDMs each containing a different third monomer. ENB = ethylidene norbornene; 1,4-HD = 1,4-hexadiene; DCPD = dicyclopentadiene.

other elastomers, the crosslinking mechanism of EPDMs with sulphur and accelerators has been researched without, however, reaching a deep understanding of the mechanism.[14]

There has been a considerable adaptation of the various accelerators offered on the market so as to obtain crosslinking conditions which would be more suited to various applications. Table 7 gives some examples.

TABLE 8
EPDM CROSSLINKED WITH RESIN, QUINOID AND DIMALEIMIDE SYSTEMS

Dutral TER 048		100	
Zinc oxide		5	
Stearic acid		1	
HAF		50	
Paraffinic oil		20	
MBTS	0·2	—	—
GMF	3	—	—
SP 1045 resin	—	3	—
Bis-methylene dimaleimide	—	—	3
Dicumyl peroxide	—	—	0·3
Press curing			
Temp., °C	180	180	180
Time, min	60	60	15
Mechanical properties			
Tensile strength, MPa	16·5	9·5	8·5
Elongation at break, %	380	660	1100
Modulus at 300%, MPa	11·2	3·4	1·3
Tension set at 200%, %	8	16	20
Hardness, Shore A	64	57	56

TABLE 9
EPDM CROSSLINKED WITH HIGH ENERGY RADIATION

Dutral TER 048	100				100				100			
FEF	80				80				80			
Paraffinic oil	40				40				40			
Ethylene glycol dimethacrylate	—				10				—			
Liquid polybutadiene	—				—				10			
Dose, Mrad	0	10	20	40	0	10	20	40	0	10	20	40
Mechanical properties												
Tensile strength, MPa	0·3	3·0	8·0	9·0	0·3	8·0	12·4	11·0	0·3	3·3	8·1	10·8
Elongation at break, %	160	600	450	360	80	450	320	230	155	710	470	430
Modulus at 200%, MPa	—	1·7	4·3	5·8	—	4·0	11·7	10·4	—	1·6	4·0	5·7
Hardness, Shore A	51	54	55	57	46	60	64	64	49	53	54	56

5.2.3. Resin, Quinoid and Maleimide Systems

Crosslinking of EPDMs with resins, quinones and maleimides is possible,[15] and final properties and heat stability are satisfactory. Table 8 shows a few examples, although in fact these systems have not been investigated and improved as the properties of EPDMs crosslinked with other systems are more satisfactory.

5.2.4. High Energy Radiation

EPDMs are more suitable than EPMs for high energy radiation crosslinking. A determining factor in keeping costs down is the use of activators,[16] of which trimethylpropane trimethacrylate (TMPT) and ethylene glycol dimethacrylate (EDMA) have been found the most efficient.

A few of the substances used as coagents in peroxide crosslinking are useful but they are often less efficient. The effect on properties when an EPDM was crosslinked by gamma rays, as a function of type of activator and quantity of energy, is shown in Table 9. Radiation crosslinking of elastomers in general, and of EPDMs in particular, is very promising but has not yet become part of industrial practice to any significant degree. So far we have knowledge of experimental production of low tension electric cables only. Excellent results, in terms of high production rates and good final properties of the insulating sheath, were obtained.

5.2.5. Room Temperature Crosslinking

A very simple method of crosslinking EPDM at room temperature is based on the use of cumyl hydroperoxide,[17] a typical compound is given in Table 10. This system has already been applied on an industrial scale in the production of pastes, cements and self-curing protective paints.

TABLE 10
ROOM TEMPERATURE CROSSLINKING OF EPDM

Mechanical properties after		Days			
	1	4	7	15	30
Tensile strength, MPa	0·5	3·0	5·9	10·5	11·7
Elongation at break, %	230	360	380	300	280
Modulus at 200%, MPa	0·4	2·1	3·9	7·1	11·2
Tension set at 200%, %	80	26	18	10	9
Hardness, Shore A	40	50	55	61	65

Recipe: Dutral TER 054, 100; zinc oxide, 5; FEF, 50; SRF, 50; paraffinic oil, 30; cobalt naphthenate, 0·1; DPG, 1; cumyl hydroperoxide, 7.

6. FILLERS AND OTHER ADDITIVES

EPM and EPDM elastomers are very much in line with traditional elastomers regarding the use of fillers and additives. As a general rule, it suffices to say that when peroxide crosslinking is involved, it is necessary to avoid or at least minimise the presence of acidic substances in the compounds. This is especially true with EPMs.

6.1. Carbon Blacks
These are still the most important fillers for reinforcing rubbers in general and EPMs and EPDMs are no exception. There are no particular mixing problems because carbon blacks are easily and quickly incorporated into EPMs and EPDMs.

Of particular interest with ethylene–propylene elastomers are conductive carbon blacks. This is because these polymers are often used in the manufacture of semi-conductive sheaths for medium and high tension electric cables. Because of the structure of this type of carbon black good dispersion is a little more difficult to achieve.

6.2. Mineral Fillers
Mineral fillers are being used more and more in elastomeric compounds. Properties are of course not comparable, at least in the majority of cases, to carbon black but many satisfactory mineral filled compounds can be obtained and very high loadings may be incorporated when high vulcanisate strength is not important.

6.2.1. Clays
Clays are probably second in popularity after carbon blacks in their use in the rubber industry. There are various qualities grouped under the terms soft, hard, calcined, and treated. Sometimes they are sold 'treated' (e.g. with fatty acids, oils, silanes and silicones) to improve properties. Calcined clays are particularly important in EPMs and EPDMs and they are currently used in the production of insulating compounds for low, medium and high tension cables.

6.2.2. Silicas
These are well known in the industry for the excellent mechanical properties they give to finished products and for the improved adhesion of the compound to metals and fabrics. Two to four parts of polyethylene glycol should also be added to the compound to avoid any detrimental interaction

between the silica and the crosslinking system. Because silica fillers do not produce good extrusion properties, it is advisable to use them in conjunction with clay when the compound is to be extruded.

With peroxide crosslinking, and therefore particularly in the case of EPMs, the use of silicas is more critical because of their acidity, which must be suitably neutralised.

6.2.3. Calcium Carbonate

This is the 'economical' filler *par excellence*, in ground, air floated and precipitated forms. It is used mainly as a diluent filler, although suitably treated precipitated types have a certain reinforcing effect. Because calcium carbonate is attacked by acids, even weak acids, its use should be avoided in compounds where the finished product could be in contact with these chemicals.

6.2.4. Hydrated Alumina

This filler is useful when flame resistant compounds, with low smoke emission characteristics are being produced. Because compounds loaded with alumina retain good insulation properties (from the electrical point of view) hydrated alumina has been used successfully in the production of electric cable insulation compounds where good flame retardance is required.

6.3. Coupling Agents

When speaking of ethylene–propylene elastomer, the term 'coupling agents' indicates silanes. These substances, added in suitable quantities to compounds containing mineral fillers, increase interaction with the polymer thereby improving the final mechanical properties.

Initially, silanes were used in compounds based on EPM containing calcined clay and which were intended for electrically insulating applications. It was noticed that the insulating properties were improved and stabilised, especially in a moist environment; mechanical properties were also improved. The same type of improvement was noted with hydrated alumina. The most suitable coupling agent, with peroxide crosslinking, is vinyl tri(1-methoxy-ethoxy) silane; with sulphur and accelerator crosslinking systems mercaptopropyl trimethoxy silane is recommended. Silanes have recently been made available as masterbatches and can be added directly to mineral fillers.The use of silanes has now extended to non-black compounds for applications other than cables when silicas and other types of filler (aluminium and calcium silicates) are used.

A new coupling agent which has appeared recently for ethylene–propylene elastomers is bis(3-(triethoxysilyl)-propyl)-tetrasulphane.

6.4. Plasticisers

The most suitable plasticiser for EPMs was found a long time ago to be polyalkyl benzene because of its low interaction with the peroxide used as a crosslinking agent. This plasticiser is no longer easily obtainable and paraffinic oils are currently being used. Naphthenic or, worse still, aromatic oils are not recommended because of their interaction with peroxides. The same comments apply to peroxide crosslinked EPDMs but if they are crosslinked with sulphur and accelerators then paraffinic and naphthenic oils are preferable. A characteristic peculiar to certain types of ethylene–propylene elastomers, particularly the higher molecular weight types, is the ability to accept large quantities of oil in their compounds, e.g. 100 parts, and more, per 100 parts of polymer. The oil must, however, be of good quality with low polar impurity content and low volatility.

A very useful plasticiser has been found in liquid polybutadiene. In peroxide crosslinking this substance also has the advantage of acting as a coagent and thus imparts improved mechanical properties to the finished products. The following substances can also be used as processing aids:

1. Low density, low molecular weight polyethylene
2. Polyethylene waxes
3. Fatty acids
4. Polyglycols

Apart from polyethylene, these are all substances which are not compatible with ethylene–propylene elastomers, and on reaching the surface they act as lubricants between the compound and the surfaces of the machine with which they are in contact, thereby improving flowability. Because of their incompatibility, they should be used in small quantities.

6.5. Tackifiers

Very probably the main cause of EPDMs not being used in the manufacture of tyres is their intrinsic lack of building tack. Tack is the ability of two surfaces of rubber or raw compound to adhere to each other when placed together under slight pressure. Lack of tack is common to almost all synthetic rubbers. Without thinking specifically of tyres, where a high degree of building tack is indispensable, there are many other applications where different layers of raw compound must be plied together. Examples are conveyor belts, medium and large fabric covered hose, etc. In these cases, a certain degree of tack is necessary when building the article. Special low molecular weight resins, generally phenol based, have been used recently as tackifiers. The quality used must be chosen carefully because they can interfere with the whole of the compound causing a general deterioration of the properties of the cured product.[18]

It has been observed that the molecular weight distribution, as well as the average molecular weight and the flexibility of the molecules (dependent on the ratio of bonded ethylene to propylene) are correlated to the degree of tack.

6.6. Anti-blooming

Surface blooming on vulcanisates is a phenomenon which afflicts all elastomers. Many of the more active ingredients, such as curatives and ageing inhibitors, are generally incompatible with elastomers, particularly if of low or nil polarity and they, or their by-products, tend to migrate to the rubber surface. The non-polar EPMs and EPDMs are particularly badly afflicted by this phenomenon. Suggestions to reduce, if not eliminate blooming are of a general nature:

1. Use the smallest amount possible of polar substances (or substances which could produce polar by-products)
2. Because there is a migration limit, use small quantities of many similar polar products rather than a larger quantity of a single substance. (This could be the case with accelerators)
3. If possible, the polar substances should be of high molecular weight
4. Polar substances which migrate easily should be blocked with other high molecular weight polar substances, such as polyethylene glycol. This technique is widely and successfully used with ethylene–propylene elastomers particularly those crosslinked with peroxides

In certain cases blooming can be accelerated and practically removed by treating the articles in a steam autoclave.

6.7. Antioxidants

It is known that peroxide crosslinking of EPDMs, and of course of EPMs, will produce articles with good resistance to high temperatures; this resistance can, however, be further improved if suitable antioxidants are added to the compound. The choice of suitable antioxidants has in fact been very difficult because of the interaction these ingredients generally have with the peroxides used for crosslinking. Polymerised dihydroquinoline is currently used in the ratio of 0·5 to 1 phr and, particularly in non-black formulations, this antioxidant is often used in conjunction with mercaptobenzimidazole. The presence of zinc oxide is always necessary, despite the fact that its function is not always clear. An antioxidant based on a blend of alkylated phenols is less staining but also less efficacious.

In sulphur/accelerator crosslinked EPDMs, only nickel butyl dithiocarbamate has been found to be effective as an antioxidant. The by-products of the various cure systems probably also act as antioxidants, but experience has shown that it is not necessary to protect EPM and EPDM based compounds when service temperatures do not exceed 100 °C.

6.8. Anti-UV

As stated earlier ethylene–propylene elastomers are sensitive to light and UV rays. If the compound is black then the carbon black will act as an absorber and can protect articles from the effects of daylight exposure for decades.

In the case of non-black compounds, experience suggests the following recommendations if good UV resistance is to be obtained:

1. Use high molecular weight EPMs and EPDMs
2. Use paraffinic oils of great purity
3. Add 15–30 phr of rutile titanium dioxide to the compound together with very small quantities of SRF carbon black (if acceptable)
4. Use phthalocyanine type pigments

The various anti-UV substances currently being used with plastics are not particularly efficient in EPM and EPDM.

6.9. Flame Retardants

The combination of chlorinated paraffin wax and antimony oxide is well known as a flame retardant, but unfortunately it evolves fumes and very aggressive chlorinated compounds.

To avoid the formation of fumes, the use of hydrated alumina, as already mentioned, together with phosphoric esters is recommended.

7. PROCESSING

7.1. Mixing

Difficulties can be encountered when use is made of high green strength EPMs and EPDMs to obtain low viscosity compounds. The problem derives mainly from the need of longer mixing times so as to obtain good dispersion of ingredients in the compound. These difficulties can be easily overcome by using elastomers in the form of soft friable bales (apparent density 0·5) instead of compact bales.

The same results can be obtained by using polymer forms such as pellets, crumbs or granulated bales. The results of mixing time comparisons are shown in Fig. 10.[8]

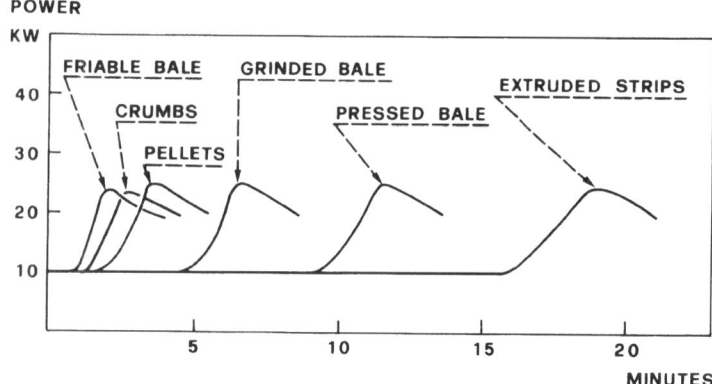

FIG. 10. High green strength EPDM (Dutral TER 038) mixing test—power absorption.

The advantages of the particulate form are as follows:

1. Easy handling
2. Automatic weighing facility
3. Continuous mixing machine feed
4. Shorter mixing cycles
5. Less energy consumption for the mixing process
6. Better dispersion in the compound
7. Lower capital investment (for new plants)
8. Feasibility of automating the whole production cycle

The disadvantages are:

1. Higher elastomer production costs and therefore a higher price of the elastomer
2. Higher packing costs because of the greater volume
3. High haulage costs (again because of the greater volume)
4. New investment required to install suitable equipment for handling

Some ethylene–propylene elastomer manufacturers already have on the market a selection of high green strength types in particulate form. Speaking in general, the upside-down mixing technique is now widely adopted because it has been found to be the most suitable, particularly when higher filler loadings are employed. High viscosity polymers (i.e. 60 and over on ML (1 + 4) 125 °C) are always difficult to work on open mills and an internal mixer is essential when the compound is being made.

7.2. Shaping
After mixing, the compound must take on the shape of the required article. It is here that the rheological behaviour of the polymer becomes important. Past experience shows that this depends mainly on the following parameters:

1. Molecular weight
2. Molecular weight distribution
3. Composition (basically the ethylene/propylene ratio)

7.2.1. Handling
This characteristic is basically connected to the green strength of the compound which depends on the green strength of the base polymer and this in turn depends almost entirely on the propylene content. Only EPMs and EPDMs with a propylene content around 30 % have a green strength which is suitable for automatic handling during building processes.

High molecular weights, with a wide molecular weight distribution, can increase the green strength of high propylene content polymers. The values reached will not be the same as those of low propylene content polymers, but the results will be satisfactory for most processes. Furthermore, green strength achieved by using higher molecular weight with a wide distribution is less temperature dependent. Green strength due to a lower propylene content will not noticeably decrease at around 50–70 °C.

7.2.2. Extrusion
A good extrusion compound must be easy to handle, have a fast extrusion speed, no (or very little) swelling and show stability of extrudate shape even at curing temperatures.

With ethylene–propylene elastomers as with all other types of elastomer, a compromise must be reached, as some of these requirements, particularly low swelling and shape stability, are contradictory.

Shape stability can be improved by using rubbers with higher molecular weights and wide distribution, but this will increase swelling. Swelling also increases with extrusion speed but can be reduced by the use of compounds containing high loadings of medium reinforcing fillers. This means that all types of ethylene–propylene elastomer available on the market, even the higher viscosity types which contain 100 parts of oil per 100 parts of elastomer, are capable of producing easily extruded compounds. Basically, it rests on the ability of the technologist to choose a polymer, or a blend of different types, which can be adapted by a choice of suitable quantities of filler to meet the particular processing and application requirements. Obviously, when a superior quality compound is required, a medium–low molecular weight elastomer will be chosen, with medium MWD and high propylene content. The higher molecular weight and green strength polymers are, however, particularly suitable, because of their filler loading capacity, for good extrusion giving satisfactory mechanical properties and a very competitive cost.

Extruders are generally fed with cold, or with only slightly heated, compounds as otherwise overflow phenomena may occur in the feeding zone. This problem can be reduced by using blends of polymers with very different intrinsic characteristics, i.e. different in molecular weight, or in MWD or in green strength.

7.2.3. Calendering

The calendering properties of EPM and EPDM based compounds, particularly those with a high filler content, is excellent.

There has been considerable interest in recent years in calendered sheets which are to be used in the uncured state (see Section 8.5). In this case, the use of EPMs and EPDMs with a very high green strength is imperative. The addition of low density polyethylene further improves the calendering properties of the compound and the tenacity of the finished sheet.

7.2.4. Moulding

In general the moulding of ethylene–propylene based compounds presents few difficulties even when complex moulds are employed. Precautions should be taken with peroxide vulcanisations so as to avoid air being trapped in the mould and causing sticky surfaces. Such precautions consist mainly of supplying suitable vents in the more critical parts of the mould.

Shrinkage is of the same order of magnitude as with other elastomers and is caused by thermal expansion which is greater than that of steel. It can be reduced by decreasing the quantity of polymer in the compound, i.e. increasing the filler content.

7.3. Curing

There is no limitation on the curing of EPDMs as all the methods currently adopted by the rubber industry can be used. For example, moulding in various ways, by rotocure, by hot air, in molten salts, in steam at various pressures, in fluid beds, by UHF high energy and by radiation.

TABLE 11
CLEAR EPDM COMPOUND FOR UHF CURING

Dutral TER 046/E3	100
Stearic acid	1
Zinc oxide	20
Titanium oxide	10
Sillikolloid	220
Polyethylene glycol (Carbowax 4000)	10
Mercapto silane (A. 189)	1
Paraffinic oil	90
Calcium oxide (Caloxol W–3)	10
MBT	1·5
ZD_mC	1·5
DPTTS	1
TMTD	1
END–75	0·7
Sulphur	2
Mechanical properties after UHF curing	
Tensile strength, MPa	6·5
Elongation at break, %	350
Modulus at 200%, MPa	4·5
Hardness, Shore A	52

Ethylene–propylene elastomers have considerable economic advantages based on their capacity of being cured at high temperatures, thus considerably reducing production times. Particularly with EPMs (with a peroxide cure system) it is possible to use a temperature as high as 250 °C when curing in molten salts, with no appreciable reduction in mechanical properties of the vulcanisate (Fig. 11). Consequently, EPMs and EPDMs are not affected by the drawbacks of over curing. As already mentioned, further reductions in cure times are obtainable using EPDMs with a high third monomer content.

One of the more interesting developments in the curing of EPDMs was recently obtained with non-black compounds cured with UHF equipment. A suitable formulation is given in Table 11, and this has been widely tested and evaluated on an industrial scale.

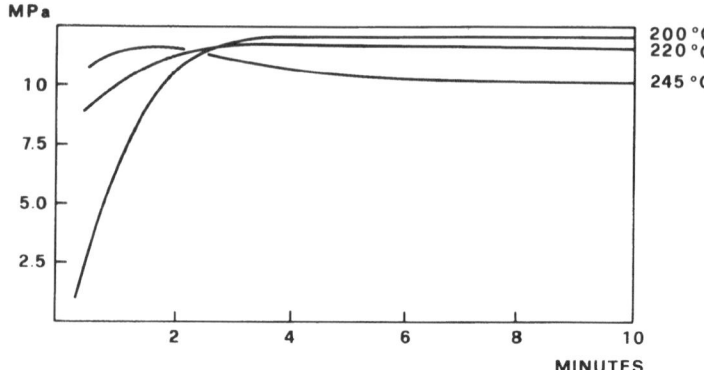

FIG. 11. Tensile strength of EPM (Dutral CO/054) compounds cured in molten salts at different temperatures.

8. APPLICATIONS

The areas of application where the use of EPMs and EPDMs is now well accepted are numerous. Table 12 shows the main areas, indicating also the relative consumption percentage, on a very approximate basis— approximate because of the difficulty of obtaining precise data.

The major markets for ethylene–propylene rubbers are those which call for high quality and long-term retention of properties.

The following comments relate to the position for the major applications.

TABLE 12
EPM AND EPDM APPLICATION FIELDS

Field of application	%
Automotive	42
Electrical cable	15
Building	10
Appliance	10
Industrial	9
Blends with plastics	9
Other	5
	100

8.1. Automotive Applications

8.1.1. Tyres

The addition of EPDM to compounds for tyre sidewalls to improve their ozone resistance was at first limited to the types containing dicyclopentadiene. Recently, and with equal success, a few ENB types have been used, such grades also having a low propylene content and medium–high unsaturation. In addition to being used in white compounds, EPDM is now also used in black compounds.

A more recent application involves inner tubes which currently are made of compounds based on butyl rubber but which have the following disadvantages: high 'cold flow' and low green strength in the uncured state, large losses by hysteresis, low resilience and high permanent set in service.

The addition of 15–20 phr of medium unsaturation EPDM will reduce these defects and will considerably improve low- and high-temperature behaviour.[19] Furthermore, the addition of 10–15 phr of EPDM to butyl rubber compounds, used to produce curing bags, will considerably increase their service life.

8.1.2. Technical Products

It is in this field of application that EPMs and EPDMs have found their main market. It is easier in this case to list the articles which cannot be produced with these elastomers, i.e. those which are directly in contact with fuels and lubricating oils.

The energy crisis of recent years, coupled with greater safety criteria, has led to the substitution of bodywork parts and of the traditional metal bumpers with impact resistant plastics or vulcanised rubber parts. Here again, EPMs and EPDMs are the preferred elastomers as they can also be used as modifiers for the more commonly used plastics (see Section 8.8). One of the more important technological breakthroughs connected with this application is the painting of EPDM based articles. The process includes UV treatment of the surfaces to be painted, using benzophenone as the activator. The paints are based on polyurethanes. All the steps are automated and perfect colour matches can be made with metal parts.[20]

8.2. Electrical Applications

8.2.1. Cables

From their beginning, ethylene–propylene elastomers looked promising as insulating materials, and in fact they were used for medium tension cable insulation before they were used for low tension cables. Their use in low tension cables has been consolidated, progressively taking over together with polyethylene, all other materials, EPMs are now also being accepted in the field of high tension cables. Their increasing success is due to their lightness compared to the traditional cables insulated with oil impregnated paper, to their relative simplicity of installation, and mainly to the considerable progress made in construction techniques which has greatly increased their reliability.

TABLE 13

EPM INSULATING COMPOUND FOR MEDIUM
AND HIGH VOLTAGE CABLE

Dutral CO/054	100
Low density polyethylene	20
Zinc oxide	5
Stearic acid	1
Antioxidant	1·5
Calcinated clay	80
Paraffinic oil	20
Vinyl silane (A–172)	1·5
Liquid polybutadiene	6
Peroxide (Peroximon F 40)	4
Mechanical properties after curing at 165°C for 40 min	
Tensile strength, MPa	6·0
Elongation at break, %	370
Modulus at 300%, MPa	5·5
Hardness, Shore A	67

Recent techniques of triple extrusion allow simultaneous extrusion of an internal semi-conductive screen, an insulating layer and an outer semi-conductive screen. In this way, practically all cavities and porosity are eliminated, thus reducing ionisation phenomena to negligible levels. Ethylene–propylene elastomer resistance to the latter is in any case very good.

Their use in electric cable insulation up to approximately 60 kV is now confirmed. Intense experimental and development work is currently under way for higher tensions and several 150 kV cables have been installed with excellent results so far. Obviously the construction techniques of cables are continuously evolving so as to increase the degree of reliability of the cables.[21,22] A typical medium and high tension formulation is given in Table 13.

One material which is used as an alternative to ethylene–propylene elastomers is polyethylene (crosslinked or otherwise). Its use, however, requires more sophisticated equipment and greater operator ability if porosity or blistering, which could seriously undermine the functioning of the cable, is to be avoided.

Ethylene–propylene elastomers are being more and more widely used in the construction of electric cable terminals. This is an application which is not well known but the application of ethylene–propylene elastomers in this application is prestigious because many assembly, maintenance and reliability problems have been solved. Basically, these are insulating compounds which are very similar to those used for medium and high tension cable insulation and which are either compression, transfer or injection moulded to produce these important accessories.

8.2.2. Insulators

This is a recent but very promising application. In this case, the EPM based compounds are insulating, but they contain hydrated alumina as a filler which gives the compound greater tracking resistance. The compound is used to mould covers for epoxy resin rods reinforced with glass fibre. This type of insulator is very reliable, unbreakable, and weighs 10 times less than its porcelain counterpart.

Energy saving field transmission lines are being planned to carry very high tensions of up to 1000 kV, and only this type of insulator is usable because the heavy porcelain type would require excessively large pylons to support them.

8.3. Building Applications

The building industry requires long lasting materials and products, at competitive costs. Ethylene–propylene elastomers are ideal materials because of their intrinsic properties. A wider range of products is employed by the building industry: door and window seals; waterproofing sheets (both cured and uncured); bearing pads for reinforced concrete structures and expansion joints for bridges and viaducts; boat and dock fenders; quality industrial flooring and sealing mastics. Interesting results have recently been obtained with EPMs in a blend with bitumen to improve the low- and high-temperature behaviour of the latter when used in road surfacing or for uncured waterproofing sheets.

More recent is the use of EPMs and EPDMs as components of solar panels. These polymers can guarantee long life at a reasonable price. EPDMs have been used in the construction of the active part of such panels,[23] the heat absorber consisting of a series of EPDM based tubes which are extruded simultaneously and subsequently cured. Inside these tubes runs water which is the heat exchange medium. It is possible to manufacture panels of any length with simple extrusion techniques.

8.4. Domestic Appliances

These are mainly items for washing machines, dish washers and driers. Compared to traditional elastomers, EPM and EPDM elastomers have considerable advantages. These include better ageing properties, high temperature resistance and the absence of the well known staining phenomena of enamelled and painted surfaces. Where these polymers are used there is no need for staining substances such as antioxidants and antiozonants to be incorporated. EPM and EPDM elastomers are also to be preferred for their greater resistance to detergent and bleaching solutions.

In the case of non-black compounds, particular precautions should be taken in the choice of mineral fillers, avoiding those which could be attacked by detergents.

8.5. Agriculture

The use of waterproof sheets based on EPDMs and EPMs for the construction of water reservoirs of any type is a widely known application. For practical as well as economic reasons, preference is given to sheets used in the uncured state as such sheets are based on high green strength ethylene–propylene elastomers blended with plastics. This application is expanding rapidly due to the growing need to recover and treat water. Another promising application in advanced stages of development is the use of EPM and EPDM based pipes for controlled underground irrigation. The pipes have regularly spaced slits which release water under controlled conditions depending on the pressure in the pipeline. Compounds to produce this type of pipe must obviously have a well defined and, if possible, constant elastic modulus together with good tear resistance so that pressure peaks can be withstood.

The good results obtained so far in this application with ethylene–propylene elastomers are also due to their inertness to bacteria.

8.6. Chemical and Mechanical Industries

In this field, the uses of EPMs and EPDMs are very numerous: conveyer belts, normal and reinforced pipes with various inserts, sealing gaskets, roller covers, tank linings, flooring, additives for oils, waxes and paints, etc.

This is the typical field of application where good adhesion is required to a variety of other materials such as textiles of various types, metals, other polymers, etc.

Although ethylene–propylene elastomers are not the best rubbers for obtaining a good degree of adhesion with other materials, satisfactory results have been obtained with the use of suitable adhesives or intermediates. Considerable research work has been carried out on the modification of EPDMs by grafting on more reactive functional groups.[24–26] Satisfactory results have been obtained but the cost of the operation has discouraged commercial development of such products.

EPM or EPDM based conveyor belts are now manufactured without difficulty and are used out of doors or in severe service conditions up to continuous temperatures of 120 °C with peaks of 140–150 °C.

For some years, in competition with other polymers, EPMs have been used as additives for lubricating oils; becoming what is known as oil viscosity improvers by increasing the high-temperature viscosity and reducing the low-temperature viscosity. Compared to other polymers, EPMs are more stable to shear forces and do not leave carbonaceous deposits. The most suitable have been found to be those with low molecular weight and narrow MWD.

Ethylene–propylene elastomers are also used as additives (10–15 %) for waxes used in paper coating. EPMs are preferred because of the absence of gel and greater UV stability, and because they confer greater self-adhesion of the paper and higher impermeability. In direct competition to this type of paper are polyethylene and polypropylene film.

8.7. Blends with Other Elastomers

The first blends of EPM and EPDMs with traditional unsaturated elastomers were made with the intention of giving the latter all or part of the exceptional ozone resistance of the former. There were difficulties, however, caused by incompatibility of the polymers themselves and lack of suitable cure systems. The incompatibility of the polymers caused problems in achieving a homogeneous mix; the lack of cure systems that were compatible and soluble in the various polymers, produced a final compound with poor mechanical properties as shown in Fig. 12. Development has been carried out to improve the compatibility of blends by, for example, halogenating EPDM, introducing other chemically active groups and seeking a third monomer and vulcanising systems[26–31] which will give acceptable cure characteristics.

It has been established that the best protection of unsaturated rubber against ozone is obtained by using 15–30 % very high unsaturation EPDM with a high ethylene content and high molecular weight. Unfortunately, good homogeneity of the compound is normally only obtained when the elastomers to be blended are of the same viscosity at the mixing

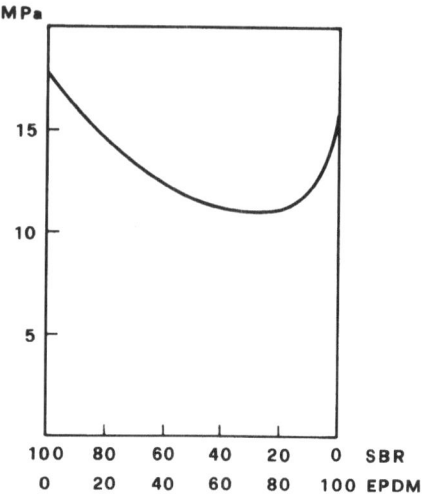

FIG. 12.   Tensile strength of normal SBR–EPDM blend based compounds.

TABLE 14a
NATURAL RUBBER/EPDM BLEND

Natural rubber (CV 5)	70
Dutral TER 046/E3	30
Zinc oxide	5
Stearic acid	2
Structol 60 NS	5
Paraffinic oil	10
Titanium oxide	5
Carbowax 4000	2
Hi-Sil 233	30
Deovulc EGL	3
Sulphur	1
Press curing: 20 min at 160°C	
Mechanical properties	
Tensile strength, MPa	12·0
Elongation at break, %	700
Modulus at 300%, MPa	2·1
Hardness, Shore A	44
Ozone resistance	
Temperature: 50°C	
Time: 70 h	
Elongation: 30%	No cracks
Ozone conc.: 50 pphm	

TABLE 14b
SBR/EPDM BLEND

SBR 1502	70
Dutral TER 046/E3	30
Zinc oxide	5
Stearic acid	1
FEF	40
SRF	40
Paraffinic oil	20
DOTG	2·5
MBT	1
CBS	0·3
Sulphur	1·5
Press curing: 20 min at 160°C	
Mechanical properties	
Tensile strength, MPa	15·0
Elongation at break, %	360
Modulus at 300%, MPa	13·5
Hardness, Shore A	65
Ozone resistance	
Temperature: 25°C ⎫	
Time: 70 h ⎬	No cracks
Elongation: 10% ⎪	
Ozone conc.: 50 pphm ⎭	

temperature. As the most suitable EPDMs must have a high molecular weight and, therefore high viscosity, they must be oil extended to obtain the appropriate viscosity. Table 14a–c gives an example of blends of unsaturated rubber with EPDM.

As mentioned in Section 8.1.1, ethylene–propylene elastomers are currently used in compounds for tyre sidewalls. In this application also ternary blends have been used, including halogenated butyl rubber.[32] With regard to EPDM and butyl rubber blends, there are no covulcanisation problems because both have a similar level of unsaturation.

8.8. Blends with Plastics

Blends of EPMs and polyethylene or polypropylene were made when these elastomers first appeared. Only in recent years, however, have they met with considerable technological and commercial interest. EPMs can be easily blended with polyethylene and/or polypropylene in all ratios, giving compounds with very interesting properties. If the elastomer is predominant, the blends thus obtained have elastomeric properties and can be used in the uncured state in many applications. In this case, we are in fact

TABLE 14c

POLYNORBORNENE/EPDM BLEND

Norsorex (polymer + oil)	175
Dutral TER 046/E3	30
Structol 60 NS	5
Zinc oxide	5
Stearic acid	1
MT–FF	100
Naphthenic	45
TMTDS	1·5
DPTTS	1·5
TDEDC	0·8
DOTG	1
DETU	0·5
DTDM	1·5
Press curing: 20 min at 160°C	
Mechanical properties	
Tensile strength, MPa	5·5
Elongation at break, %	170
Modulus at 100%, MPa	3·0
Hardness, Shore A	44
Ozone resistance	
Temperature: 40°C ⎫	
Time: 70 h ⎬	No cracks
Elongation: 25% ⎪	
Ozone conc.: 50 pphm ⎭	

talking of 'thermoplastic rubbers'. Their elastic properties can be improved by a process of partial crosslinking which occurs during the mixing stage.

If, on the other hand, the plastic material is predominant, interesting modifications of the latter are obtained such as improved impact resistance over a wide range of temperatures.[33-35]

One of the most recent applications of polypropylene blends with EPM is in bumpers and bodywork parts for small and medium sized cars. Other plastics whose properties improve with the addition of EPM are PVC, polystyrene and acrylonitrile–butadiene–styrene resins, such blends with plastic materials have promising growth prospects.

REFERENCES

1. NATTA, G., MAZZANTI, G., VALVASSORI, A. and PAJARO, G., *Chimica e. Industria*, **39**, 1957, 733.

2. NATTA, G., MAZZANTI, G., VALVASSORI, A. and SARTORI, G., *Chimica e. Industria*, **40**, 1958, 717.
3. NATTA, G., MAZZANTI, G., VALVASSORI, A. and SARTORI, G., *Chimica e. Industria*, **40**, 1958, 896.
4. CORBELLI, L., MILANI, F. and FABBRI, R., *International Rubber Meeting, Nürnberg*, 1980.
5. NATTA, G., MAZZANTI, G., LONGI, P. and BERNARDINI, F., *Chimica e. Industria*, **41**, 1959, 519.
6. NATTA, G., PASQUON, I. and GIUFFRE', L., *Chimica e. Industria*, **43**, 1961, 871.
7. CRESPI, G. and DI DRUSCO, G., *Hydrocarbon Processing*, **48**(2), 1969, 103.
8. CORBELLI, L. and NOVI, A., *Kautschuk u. Gummi Kunststoffe*, **32**, 1979, 957.
9. NATTA, G., SARTORI, G., VALVASSORI, A., MAZZANTI, G. and CRESPI, G., *Hydrocarbon Processing*, **41**(9), 1962, 261.
10. MAURER, J., *Rubber Chem. Technol.*, **38**, 1965, 979.
11. NATTA, G., CRESPI, G., VALVASSORI, A. and SARTORI, G., *Rubber Chem. Technol.*, **36**, 1963, 1583.
12. LOAN, L., *Rubber Chem. Technol.*, **40**, 1967, 149.
13. DI GIULIO, E. and BALLINI, G., *Kautschuk u. Gummi Kunststoffe*, **15**(1), 1962, 6.
14. FUJIMOTO, K. and NAKADE, S., *J. App. Polym. Sci.*, **13**, 1969, 1509.
15. GILLER, A., *Kautschuk u. Gummi Kunststoffe*, **17**, 1964, 174.
16. ELDRED, R., *Rubber Chem. Technol.*, **47**, 1974, 924.
17. CORBELLI, L. and GIOVANARDI, S., *Swedish Rubber Conference, Runneby*, 1975.
18. PETERSON, K. and SULLIVAN, J., *Rubber Age*, **108**(2), 1976, 27.
19. ZUEVA, N., GAVRILOV, V. and SAPRANOU, V., *International Polym. Sci. Technol.*, **3**(4), 1976, T/45.
20. BANKS, S., SPENADEL, L., DE PIERRI, W. and KRUSE, D., *Rubber Age*, **106**(12), 1974, 39.
21. BLODGETT, R., *Rubber Chem. Technol.*, **52**, 1979, 410.
22. LOMBARDI, G., VALLAURI, A. and VALLAURI, U., *T.E. International*, **2**(4), 1978, 25.
23. ANGIOLETTI, A., *International Rubber Conference, Venice*, 1979.
24. NATTA, G., SEVERINI, F., PEGORARO, M. and CRUGNOLA, A., *Chimica e. Industria*, **47**, 1965, 1176.
25. NATTA, G., PEGORARO, M., SEVERINI, F. and AURELIO, G., *Chimica e. Industria*, **50**, 1968, 18.
26. PELLON, J. and VALAN, K., *J. App. Polym. Sci.*, **9**, 1965, 2955.
27. MORRISSEY, R., *Rubber Chem. Technol.*, **44**, 1971, 1025.
28. MORRISSEY, R., *Rubber Chem. Technol.*, **49**, 1976, 353.
29. BARANWAL, K. and SON, P., *Rubber Chem. Technol.*, **47**, 1974, 88.
30. HOPPER, R., *Rubber Chem. Technol.*, **49**, 1976, 341.
31. MASTROMATTEO, R., MITCHELL, J. and BRETT, T., *Rubber Chem. Technol.*, **44**, 1971, 1065.
32. BLACKSHAW, G. and KRISTENSEN, I., *Rubber Age*, **107**(5), 1975, 57.
33. DANESI, S. and PORTER, R., *Polymer*, **19**, 1978, 448.
34. WEAVER, E., *Elastomerics*, **110**(5), 1978, 21.
35. DANESI, S. and BALZANI, L., *International Rubber Conference, Venice*, 1979.

Chapter 5

DEVELOPMENTS WITH POLYCHLOROPRENE

J. C. Bament and J. G. Pillow

Du Pont (UK) Ltd, Hemel Hempstead, Hertfordshire, UK

SUMMARY

Developments in polychloroprene since 1939 are briefly reviewed in chronological order. Recent manufacturing trends are discussed, particularly the switch to butadiene in place of acetylene as feed stock. The latest developments in polychloroprene including a new sulphur modified type and the introduction of superior processing polymers are reviewed. Alternatives to ethylene thiourea as the vulcanisation accelerator for polychloroprene are proposed. The selection of the correct type of polymer for a given application is discussed, also compounding to develop specific properties in polychloroprene compounds. Developments in processing including continuous curing of extrusions, injection moulding and powder mixing are reviewed.

1. INTRODUCTION

Polychloroprene or neoprene was one of the first synthetic rubbers available to the rubber industry. It now represents 5 % of the world's consumption of rubber and is one of the more important speciality elastomers due to its resistance to lubricating oils and greases, oxygen, ozone, weathering, heat and flame. It has been the subject of continuous modification to meet changing market conditions, the main developments being, in chronological order:

1939, Neoprene GN:[1] A copolymer with sulphur, polymerisation being taken to a high conversion level (91 %). Thiuram disulphide is used to peptise the copolymer to modify the viscosity for satisfactory processing and to stabilise the polymer.[2]

1944, Neoprene GN-A: Improved polymer stability obtained by adding an aromatic naphthylamine as antioxidant.

1949, Neoprene W: Based on the use of mercaptans[3] as chain transfer agents to obtain soluble homopolymers at 65–70 % conversion level which are sulphur free thus giving improved heat resistance.

1952, Neoprene WRT, GRT: Copolymers with 2,3-dichloro-1,3-butadiene[4] with reduced linearity in the polymer chain thus reducing the tendency of the homopolymer to crystallise rapidly.

1959, Neoprene WB: A homopolymer containing a high level of precrosslinking or gel polymer[5] to improve the processability, particularly reduced nerve and die swell.

1970, Neoprene T-types: Contain medium levels of gel and are superior processing versions of Neoprene W, WHV 100 and WRT but possess equivalent physical properties.

1977, Neoprene GW: A sulphur modified polymer in which polymerisation has been controlled to combine the optimum properties of both sulphur and mercaptan modification.[6] Du Pont remained the major producer of CR until 1960 when Bayer brought a plant on stream. This was followed in 1962 by Denki Kagaku, in 1963 by Showa and in 1966 by Distugil using the butadiene route developed by the Distillers Co. in England.

2. POLYMER MANUFACTURE

2.1. Acetylene Route

At the time of the discovery of polychloroprene, acetylene appeared to be the only practical feed stock since no economic route through butadiene existed. Acetylene generated by the hydrolysis of calcium carbide was commercially available, hence for the next three decades all production was based on acetylene technology represented as follows:[7]

$$2\ CH\equiv CH \xrightarrow{\text{CuCl}_2} HC\equiv C-CH=CH_2$$

Acetylene   Monovinyl acetylene

$$HC\equiv C-CH=CH_2 + HCl \xrightarrow{\text{CuCl}_2} CH_2=C-CH=CH_2$$
$$\qquad\qquad\qquad\qquad\qquad\qquad\qquad | $$
$$\qquad\qquad\qquad\qquad\qquad\qquad\qquad Cl$$

2-Chloro-1,3-butadiene

$$CH_2=C-CH=CH_2 \xrightarrow{\text{Polymerise}} (-CH_2-C=CH-CH_2-)_n$$
$$\quad\ \ |\qquad\qquad\qquad\qquad\qquad\qquad\qquad |$$
$$\quad\ \ Cl\qquad\qquad\qquad\qquad\qquad\qquad\ Cl$$

2.2. Butadiene Route

In the 1960s a major change in monomer technology began. This was brought about by the increasing cost of acetylene relative to butadiene and was accelerated by the development of lower investment processes based on butadiene by the Distillers Co. in England.

Further impetus was given by the major explosion at the Du Pont, Louisville site in 1965. The overall aspects of economy and safety have resulted in all manufacturing facilities being converted to the butadiene route. The essential steps in the process may be represented as[8]

Chlorination

$$CH_2{=}CH{-}CH{=}CH_2 + Cl_2 \xrightarrow{400\,°C} ClCH_2{-}CH{=}CH{-}CH_2Cl$$

Isomerisation

$$ClCH_2{-}CH{=}CH{-}CH_2Cl \xrightarrow{isomerise} ClCH_2{-}CHClCH{=}CH_2$$

Dehydrohalogenation

$$ClCH_2CHClCH{=}CH_2 \xrightarrow[NaOH]{Aqueous} CH_2{=}CH{-}CCl{=}CH_2 + NaCl + H_2O$$

Direct vapour phase chlorination of butadiene gives high yields (85–95 %) of mixed dichlorobutenes, which are fractionated in the presence of cupric salts to take off the 3,4-dichlorobutene which in turn is reacted with alkali to give chloroprene. Polymerisation is carried out in an oil-in-water emulsion system.[9] An aqueous solution of potassium persulphate is used as the free-radical source to initiate and control polymerisation at the selected temperature, the degree of conversion being controlled by density. With mercaptan modification, conversion is controlled to about 70 % to give good physical properties and adequate processing. With sulphur modification, conversion is taken to about 90 % and thiuram disulphide is then added to break the sulphur linkages to obtain a processable rubber and at the same time to stabilise the polymer. The isolation of the polymer is obtained by cooling the latex, acidifying with 10 % acetic acid to a pH of 6 which destabilises the latex sufficiently to allow coagulation as a thin film on a rotating, chilled, stainless steel roll partly immersed in the latex. After washing and thawing, the rubber film is dried in hot air (120 °C) and is subsequently twisted into rope and then chopped into chips.

3. NEW GRADES OF POLYCHLOROPRENE

3.1. A Sulphur Modified Type (Neoprene GW)

Mercaptan modified polychloroprene compounds vulcanised with a thiourea give superior heat ageing and compression set compared to

TABLE 1

	Neoprene GW	Neoprene GN-A	Neoprene W	Neoprene W
Modifier	Sulphur	Sulphur	Mercaptan	Mercaptan
Ethylene thiourea	—	—	0·5	—
TMTD	—	—	0·75	—
Sulphur	—	—	—	1·0
TMTM	—	—	—	0·5
DOTG	—	—	—	0·5
Mooney viscosity 121 °C (minimum value)	29	20	27	26
Cured 20 min at 160°C				
Modulus at 100 % elongation (MPa)	4·34	3·86	3·59	3·79
Tensile strength (MPa)	19·72	18·34	19·66	21·52
Elongation (%)	370	400	350	380
Hardness (Durometer A)	67	68	63	67
Tear strength (ASTM D624 'C') (kN/m)	50	45	42	48
Compression set (D395 'B') 22 h at 100°C	36	71	25	48
De Mattia flex kc to 13 mm cut growth	120	50	5	17
Aged 7 days at 121°C				
Tensile (% retained)	87	93	91	86
Elongation (% retained)	50	37	50	40
Hardness (points change)	+9	+13	+7	+13

sulphur modified types. However, the latter are superior in hot strength, resilience and resistance to flexing and tear. Thus it is necessary to choose between these two contradictory sets of properties since the molecular weight control methods are mutually exclusive. A new sulphur modified polymer was therefore introduced in 1977 which combined most of the good properties of both sulphur and mercaptan modification.[10] This was achieved by a carefully optimised combination of sulphur level, soap system, polymerisation and peptisation reactions. The new polymer is similar to a normal sulphur modified rubber in that a tight state of cure is obtainable with magnesia and zinc oxide addition only, although a thiourea may also be used to obtain a faster cure rate if required. Vulcanisates of the new polymer possess higher tear resistance than was previously obtainable with polychloroprene; resilience, flex resistance and heat ageing are equal to a sulphur modified polymer with compression set approaching that of a mercaptan modified polymer cured with a thiourea (Table 1). The test formulation is: Polymer, 100; magnesia, 4; stearic acid, 0·5; octylated diphenylamine, 2; N774 black, 58; aromatic oil, 10; zinc oxide, 5.

During processing this new polymer shows a reduced capacity for breakdown compared to the other sulphur copolymers (see viscosity figures in Table 1). Compounds therefore tend to exhibit higher viscosity, to be less prone to roll sticking and will show reduced building tack and slightly poorer extrusion smoothness. The combination of high tear strength, flex and heat resistance has been found useful for many moulded items, for instance, thin walled bellows and boots for the sealing of constant velocity joints used in front wheel drive vehicles. The good mould flow and hot tear resistance reduces defective parts and the retention of viscosity during processing contributes to reduced air entrapment.

3.2. Superior Processing Polychloroprene
Mercaptan modified homopolymers tend to exhibit nerve and die swell in processing operations. In 1959 a major improvement in processing was obtained by the addition of a precrosslinked polymer or gel. The gel polymer (Neoprene WB) possesses very low nerve and is normally used in a blend with other polychloroprenes to reduce die swell and to improve the collapse resistance of extrusions. Vulcanisates are equivalent in heat, ozone, oil and compression set resistance but have reduced tensile strength, tear resistance and flex cut growth which can be improved to a degree by a sulphur based curing system but with some loss in heat and set resistance. In view of these deficiencies, there was a requirement for a new class of polymer combining the superior processing given by the presence of a gel but with

the physical properties, particularly tensile strength, of the mercaptan modified homopolymer and copolymer. This was accomplished in 1970 with the introduction of the Neoprene T-types having a medium level of gel and normally used at 100 % of the polychloroprene content. They are available as homopolymers (Neoprene TW and TW100),[11] the latter being a higher viscosity polymer, as well as a copolymer with dichlorobutadiene (Neoprene TRT).[11] Compounds based on the T-types mix faster, developing less heat during mixing and extrude and calender on average 30 % faster than similar compounds based on non-gel-containing polymers. Die swell and shrinkage are lower and collapse resistance and die definition are also better. Despite the gel content, the mechanical properties of vulcanisates are equivalent to the W-types. The choice of a superior processing type of polychloroprene is usually made on the basis that a single polymer will mix more uniformly and give less batch to batch variability than a physical blend of two polymers.

4. CURE SYSTEMS

4.1. Ethylene Thiourea and Toxicity

Ethylene thiourea (ETU) has been used as an accelerator for polychloroprene compounds for upwards of 30 years and still remains the preferred curative for optimum compression set and heat ageing in mercaptan modified polymers. In 1969 the FDA administration in the USA found that when ETU was administered in the diet of rats at a dosage of 250 ppm, it induced a significant number of thyroid tumours.[12] Subsequently, it was found that when a solution of ETU was applied to the skin of pregnant rats, it induced a high proportion of foetal abnormalities in the offspring. As a result there has been an intensive search for an alternative to ETU. However, the British Rubber Manufacturers' Association through its Health Advisory Committee have advised that there was insufficient evidence to warrant a ban on the use of the chemical. However, they recommended that it should be used either in masterbatch form[13,14] or as a pelletised compound to eliminate any exposure to dust and that women of child bearing age should be protected from contact with the chemical by excluding them from areas concerned with mixing, milling and vulcanisation. Thus ETU continues to be widely used as a polychloroprene accelerator provided these appropriate safety precautions are observed.

4.2. Alternative Thioureas

As a result of the doubts which have arisen concerning the toxicity of ETU,

TABLE 2

	1 NA-22F 0·75 CBS	2 Thiate E 1 Epikote 828	3 TBTU	1 DETU	0·5 Sulphur 1 TMTM 1 DPG	1·25 Hexamine 1 TMTD 2 Carbowax 4000	1·25 Vanax NP 1 TMTD	1·25 Permalux 1 TMTD	1·5 Robac 70
Mooney viscosity ML 100°C 4 min	27	24	22	25	24	24	24	24	24
Mooney viscosity after 14 days at 38°C	31	30	26	46	27	39	28	28	29
Mooney scorch MS 121°C + 10 points (min)	24	45	45	15	45	31	29	18	41
Mooney scorch after 14 days at 38°C	20	37	45	2	45	21	16	11	34
ODR, TM100, 160°C time to optimum (min)	19	12	15	12	18	14	11	18	23
ODR, TM100, 190°C time to optimum (min)	4·3	3·3	3·9	2·9	3·95	3·7	3·5	3·55	5·1
Initial properties, cured 20 min at 160°C									
Tensile strength (MN/m²)	16·5	15·4	14·8	16·5	16·2	15·6	17·4	16	16·5
E/B (%)	300	210	240	320	350	300	330	330	340
M$_{200}$ (MN/m²)	9·3	14·2	11·3	8·5	7·7	9·3	8·7	8·6	8·1
Hardness (IRHD)	64	68	66	61	60	65	60	60	61
Tear strength (Die C) (kN/m)	37	31	36	39	39	35	38	40	35
Compression set 70 h at 100°C (cured 25 min at 160°C)	30	14	16	19	42	33	27	25	36
Aged 7 days at 100°C (% change)									
TS	−7	−4	−4	−13	−11	−10	−15	−2	−12
EB	0	+14	0	−7	−14	−10	−9	−3	−29
H	+3	+3	+3	+3	+8	+6	+7	+6	+9

considerable investigation into alternatives has been carried out. These are summarised in the remainder of Section 4 based on the properties obtained with a compound based on a mercaptan modified polymer (Table 2) with SRF carbon black loading.

Alternative thioureas are listed but these may have possible carcinogenic or teratogenic effects like ETU.

4.2.1. Trimethyl Thiourea (Thiate E)
Trimethyl thiourea (2 phr) with epoxy resin (1 phr) gives safer processing, faster cure rate, improved compression set, poorer tear resistance, lower elongation than ETU and is used when the best compression set is required.[15] The accelerator cost is higher than ETU.

4.2.2. Tributyl Thiourea
Again a fast curing, safe processing accelerator, used at the 3 phr level, giving good compression set values, also non-staining ozone resistance when used in combination with octylated diphenylamine. The accelerator cost is high.

4.2.3. Diethyl Thiourea (DETU)
This gives a very fast cure rate and is used in the continuous vulcanisation of both solid and cellular profiles,[15] particularly in the latter to obtain a sufficiently fast cure rate for good skin formation. Disadvantages arising from the use of this material are poor processing safety and a lacrimatory effect on the operator's eyes necessitating efficient ventilation on the curing unit.

It may also cause dermatitis when cured articles are handled. Physical properties of vulcanisates are equivalent to those obtained with ETU, with better compression set resistance, when 1 phr is used.

4.2.4. Diphenyl Thiourea
This is similar to DETU but gives slightly safer processing and is not lacrimatory.

4.3. Low Sulphur Based System[15]
A low sulphur based system consisting of 0·5 phr sulphur and 1 phr TMTM gives very safe processing and an equivalent cure rate to ETU. Physical properties of vulcanisates are also similar except for compression set and retention of elongation after heat ageing which are poorer due to the presence of sulphur. In applications such as hydraulic hose covers, where

compression set is not a major requirement, this system is a viable alternative to ETU.

4.4. Hexamethylene Tetramine System[15]
A composite system of hexamethylene tetramine (1·25 phr), TMTD (1 phr) and PEG (2 phr) gives equivalent processing safety to ETU with slightly faster cure rate and equivalent vulcanisate properties, except for compression set which is poorer. The disadvantage is that it is necessary to obtain a fine particle size hexamine to ensure good dispersion. It also tends to cause dermatitis when incorrectly handled.

4.5. Thiadiazine System
A non-carcinogenic thiadiazine (80 %) with 20 % dibutyl dithiocarbamate commercially known as Vanax MP (supplier Vanderbilt) has recently been introduced and provides a viable alternative to ETU when used at 1 phr in carbon black filled compounds.

4.6. Permalux System
The DOTG salt of dicatechol borate, known as Permalux (1·25 phr), with TMTD (1 phr)[15] gives poorer processing safety and bin storage stability than ETU, cure rate and vulcanisate properties being equivalent. Permalux is sensitive to the type of magnesia used and tends to instability in the presence of moisture which may be controlled by using calcium oxide as a desiccant in the compound. Additional processing safety and cure rate with improved compression set of the vulcanisate are obtained with 0·5 phr epoxy resin and 0·5 phr DOTG in place of 1 phr TMTD. With mineral filled compounds, the scorch and bin storage stability is inferior to ETU.

4.7. Robac 70
Dimethyl thiocarbamoyl 2-imidazolidenethione, known as Robac 70 (supplier Robinson Brothers), is a very safe processing accelerator for polychloroprene and decomposes during vulcanisation to give free ETU. Vulcanisate properties, when Robac 70 is used at 1·5 phr, are equivalent to those obtained with ETU, except for compression set and heat ageing which are poorer. The cure rate is slower than with ETU.

4.8. General
In considering alternatives to ETU, the newly introduced sulphur modified polymer (Neoprene GW) is a viable alternative since good vulcanisate

properties, particularly tensile strength, tear and flex resistance, may be obtained in short cure cycles at 170–180 °C without the use of any accelerator. For instance, this new polymer gives a cure rate which is fast enough for continuously vulcanised cable jackets to be cured in 1 min at 205 °C. Meanwhile, development continues in order to find a completely viable alternative to ETU. The latest candidate available is dimethyl ammonium hydrogen isophthalate, commercially known as Vanax CPA and marketed by Vanderbilt. Using 1–1·5 phr of this new accelerator gives the best balance of scorch and cure rate, vulcanisate properties being comparable to those obtained with ETU in both black and clay filled compounds. Bayer are also now marketing Vulkacit CRV as a counterpart to ETU.

5. COMPOUNDING FOR SELECTED PROPERTIES

Polychloroprene is a particularly versatile elastomer because it has a combination of properties suitable for many applications. It gives vulcanisates which have high tensile strength, resilience and abrasion resistance, and which resist deterioration by oils, solvents, weather, oxygen, ozone, heat and flame. It is used when service conditions are severe and especially where a product must withstand a combination of conditions. Each property may be enhanced by selective compounding, but it must be recognised that compounding for the optimum in any one property may lead to a sacrifice in some other. All requirements of processing and end use must be kept in mind when applying compounding techniques to improve certain chosen properties.

5.1. Selection of the Correct Type of Polychloroprene
For a given application three factors need to be considered. Firstly, the most important physical properties for optimum service life must be identified, i.e. flex resistance for V belt drives, compression set for seals, high and low temperature performance for automotive joint boots. Secondly, the degree of crystallisation resistance required for the operating temperature involved in the application or alternatively, for processing reasons, must be determined. Thirdly, the correct viscosity polymer for the processing operations involved in manufacture must be selected. Table 3 gives the basic properties of the three main groups of polychloroprene both for the raw polymers and the vulcanisates.

TABLE 3
CHARACTERISTICS OF POLYCHLOROPRENE

	Sulphur modified types	Mercaptan modified types	Gel-containing types
Raw polymer	Limited storage stability	Excellent storage stability	Excellent storage stability
	Peptisable to varying degrees	Non-peptisable	Least nerve non-peptisable
	Fast curing but safe processing		Best extrusion calendering performance
	Accelerators not always necessary	Needs acceleration	Needs acceleration
Vulcanisates	Best tear strength, best flex, best resilience	Best compression set resistance, best heat ageing	Properties similar to mercaptan modified types

5.2. Environmental Resistance Requirements

The main factors in the effects of weather are exposure to ultra-violet light and ozone with temperature and humidity being contributory factors.

Ultra-violet light must be screened out with an effective pigment to avoid the rapid surface crazing which will occur on exposure to bright sunlight. The most effective pigment is carbon black, a minimum of 15 phr furnace black (SRF) being required for complete resistance. In non-black compounds, rutile titanium dioxide (30 phr) and red or yellow iron oxide (10–15 phr) will give less complete protection.

Vegetable oils should be avoided as their presence may lead to fungal growth.

Ozone resistance is possessed to an inherent degree by polychloroprene, but the use of antiozonants and microcrystalline waxes is necessary for demanding applications.

Moderate ozone test requirements (70 h at 50 pphm, 38 °C, 20 % elongation) may be met on good quality compounds using 2 phr octylated diphenylamine with 3–5 phr of a microcrystalline wax.

Demanding requirements (100 h at 300 pphm, 38 °C, 20 % elongation) require an antiozonant in addition to the microcrystalline wax. This may be 1–2 phr of a mixed diaryl hindered p-phenylene diamine (e.g. Wingstay 100) if staining may be tolerated or 1–3 phr of one of the new powerful non-staining antiozonants such as tris(5-norbornene 2-methyl)phosphite. The use of the wax is important in order to maintain a film of the antiozonant at the surface. Care should be taken in choice of antiozonant because non-hindered p-phenylene diamines can seriously affect the bin storage stability of a compound.

Most plasticisers, especially esters, detract from ozone resistance with the exception of butyl oleate which is beneficial.

5.3. Heat Resistance (Table 4)

Heat resistance requires the use of a mercaptan modified polymer (W- or T-type) which does not contain sulphur. Sulphur and sulphur bearing materials such as factice should be avoided.

TABLE 4
HEAT RESISTANT NEOPRENE COMPOUNDS TO MEET DTD 5514 SPECIFICATION

Hardness range	45 ± 4	55 ± 4	65 ± 4	75 ± 4	85 ± 4
Neoprene WRT	60	70	80	80	80
Neoprene WB	40	30	20	20	20
Octylated diphenylamine	4	4	4	4	4
Heat resistant antioxidant	1	1	1	1	1
Calcium stearate	5	5	5	5	5
Petrolatum	1	1	1	1	1
SRF carbon black	25	—	—	—	—
FEF carbon black	—	35	45	—	—
SAF carbon black	—	—	—	35	50
Precipitated silica	—	—	—	15	25
Rapeseed oil	15	10	7·5	—	8
Dioctyl sebacate	10	10	5	8	8
Pitt Consul 500	2	—	—	—	—
NA-22 dispersion (75 %)	0·75	0·63	0·75	0·75	0·75
TMTD	—	—	0·70	0·70	0·70
Zinc oxide	5	5	5	5	5

The selection of an antioxidant is very important, the preferred one being octylated diphenylamine which may be used up to a level of 6 phr.

Volatile plasticisers such as butyl oleate and naphthenic oils should be avoided while rapeseed oil and polyester plasticisers are recommended for permanence. Fine particle size calcium carbonate confers good heat resistance but gives poor weathering performance and low physical properties. Carbon blacks such as FEF or SRF are usually needed.

The curing system should contain a high level of acid acceptance. High activity magnesia (4–6 phr) with up to 10 phr of zinc oxide and use of a thiourea accelerator is recommended.

5.4. Low Temperature Resistance

The low temperature performance of polychloroprene vulcanisates involves two separate factors:

1. Crystallisation, which occurs fastest at around $-10\,°C$ is time dependent, and is seen as a hardening and stiffening process
2. Second order transition, which occurs at temperatures approaching $-45\,°C$, is not time dependent and is seen as a change from the rubber like state to a glassy, brittle state

The most important factor in controlling crystallisation rate is selection of the polymer. The use of a copolymer with dichlorobutadiene is essential and a mercaptan modified copolymer (WRT or TRT) gives the best results. The use of aromatic oils, resinous plasticisers and sulphur cures, all help to retard crystallisation. Ester plasticisers, by contrast, should when possible, be avoided as they will accelerate the effect.

Ester plasticisers, however, are essential for lowering the second order transition temperature. Dioctyl sebacate (DOS) is commonly used because it is effective in reducing the brittle point below $-40\,°C$ while still giving adequate heat ageing. Butyl carbitol formal (BCF) and butyl oleate are more effective but are volatile at temperatures around $100\,°C$. Phthalates are not so effective but are used in less demanding conditions since they are cheaper.

Full flexibility at $-40\,°C$ requires 15 phr DOS for a 50 durometer A compound and 25 phr for a 70 durometer A compound. Table 5 shows a typical compound designed for good low temperature stiffening resistance.

5.5. Water Resistance

Good water resistance of polychloroprene vulcanisates relies on the absence of water-soluble components. Obvious ones such as calcium oxide desiccants and glycols may be omitted. However, water-soluble chlorides can be formed during vulcanisation. For this reason calcium carbonate fillers should always be avoided and magnesia/zinc oxide excluded from compounds needing maximum resistance.

Since oxides of lead form relatively insoluble chlorides, up to 20 phr of red lead may be used to replace the magnesia/zinc oxide acid acceptor, giving optimum protection. However, the use of red lead reduces the processing safety and for this reason can usually only be applied to low viscosity mercaptan modified (W) type compounds which are sulphur cured.

Normal magnesia acid acceptance systems, based on a high activity

TABLE 5

CONSTANT VELOCITY JOINT BOOT (VW639
PART A, FOR AUTOMOTIVE APPLICATION
50 ± 5 DUROMETER A)

Neoprene WHV	35
Neoprene WRT	65
High activity magnesia	4
Octylated diphenylamine	4
Heat resistant antioxidant	1
Antiozonant	2
Microcrystalline wax	2
Stearic acid	0·5
SRF carbon black	20
MT carbon black	50
Dioctyl sebacate	15
Rapeseed oil	10
NA-22 dispersion (75%)	1
TMTD	1
Zinc oxide	5

Cure: 15 min at 153 °C.

magnesia may be found satisfactory especially where less demanding needs are involved. Water from aqueous solutions, including seawater, is absorbed less than from deionised water and the absorption decreases as the strength of the solution is increased. It follows, therefore, that water absorption tests in deionised water may give results which are rarely encountered in actual service. The use of magnesia with up to 25 phr of precipitated silica is useful for long-term water resistance. Initially the water absorption is high (up to 20 days at 70 °C) but then it decreases and reaches a stable equilibrium. Silica is not recommended for use in composites such as cable jackets because it has been shown to be the cause of water pockets between insulation and jacket. Aqueous based chemical solutions require the use of red lead cure systems for maximum resistance. When acids are involved it is important to use an inert filler such as barytes or blanc fixe. Tank lining compounds are usually based on mercaptan modified (WRT) copolymers in order to get maximum tack (Table 6).

5.6. Resistance to Oils and Solvents

The swelling of polychloroprene vulcanisates by oils, chemicals and solvents may be reduced considerably by increasing the state of cure, by increasing the amount of loading, or by using a plasticiser which is miscible with the immersion fluid and consequently is extracted by it. Highly

TABLE 6

TANK LINING

Neoprene WRT	80
Neoprene WB	20
90% Red lead dispersion	22
Octylated diphenylamine	2
Stearic acid	1
SRF carbon black	50
Aromatic process oil	10
Sulphur	0·75
TMTM	1

Cure: 20 min at 153 °C.

extended WHV-type compounds have excellent resistance to swelling because of their low polymer content.

Blends with nitrile rubbers will give improved oil resistance but a poorer level of vulcanisate physical properties. They are normally limited to sulphur modified types using a cure system based on TETD (0·2–0·5 phr).

5.7. Mechanical Properties

Polychloroprene vulcanisates have excellent resistance to flex cracking. However, for products that will be flexed severely in service, compounds should be soft, with low modulus and high elongation wherever possible. This should be done using well dispersed loadings of soft carbon blacks (MT or SRF) rather than mineral fillers, although up to 20 phr precipitated silica will be beneficial. A good antioxidant/antiozonant protection system such as 2 phr octylated diphenylamine and 1 phr mixed diaryl *p*-phenylene diamine is required. The plasticiser should be kept to a minimum and preferably be an aromatic oil; esters should be avoided.

Mercaptan modified (W) types of polychloroprene are not as good as sulphur modified (G) types, the copolymer (GRT) being preferred. Gel-containing polymers are particularly poor.

Vibration damping is required in applications such as machinery mounts in an oily environment. For high damping characteristics, highly filled compounds should be used based on high viscosity mercaptan modified (WHV) types with high loadings of soft black, china clay and aromatic oil. Thiourea cure systems are required to minimise creep.

5.7.1. Abrasion Resistance

Abrasion resistance is improved as in other elastomers by the use of reinforcing type fillers. The best properties are obtained by using the fine particle size carbon blacks such as HAF, ISAF, SAF and CRF at about 40 phr level with a minimum of plasticiser. Silicas are the best of the non-black fillers.

5.7.2. Tear Strength

Greater tear strength is obtained from sulphur modified (G) type polychloroprene than the mercaptan modified types, with Neoprene GW giving the highest values. Mineral fillers such as fine particle size silicas, silicates and hard clays give higher tear values than carbon blacks but at the expense of poorer compression set. A good balance of tear and set properties can be obtained by using a low structure, low modulus and high elongation furnace black such as N326 (care must be taken to optimise dispersion of this often difficult to handle material). Resinous plasticisers such as coumarone–indene resins and in particular hydrocarbon resins such as Escorez 3102 at a level of 5 phr will give an improvement, but not in hot tear strength. If the presence of some natural rubber in a polychloroprene compound does not detract too greatly from other vulcanisate properties, 10–20 phr is particularly helpful.

5.7.3. Compression Set

Compression set resistance can be improved considerably in all types of polychloroprene by effecting a very tight cure. Thus, long cure times, high curing temperatures and high accelerator concentrations are beneficial. While acceleration with ethylene thiourea (1 phr) or diethyl thiourea (1 phr) for fast continuous vulcanisation will meet most requirements, for the most demanding specifications it may be necessary to use trimethyl thiourea (2 phr) and epoxy resin (1 phr) or tributyl thiourea (3–4 phr).

Compression set resistance is greater with carbon black loading than with mineral filler loading, the best results being from larger particle size blacks. If mineral filler loading must be used a calcined clay is comparatively good.

The mercaptan modified polychloroprenes are vastly superior to the sulphur modified (G) types in set resistance although Neoprene GW has partly closed the gap. For good set resistance at low temperatures, the crystallisation resistant copolymers must be used, with compounding appropriate to low temperature resistance.

Creep resistance has similar requirements to compression set resistance. For optimum values a mercaptan modified copolymer (WRT) should be used with carbon black reinforcement, a minimum amount of plasticiser and a tight thiourea accelerated cure.

5.8. Coloured Products

Coloured polychloroprene vulcanisates have limited application because of their tendency to discolour, especially when exposed to strong sunlight. In cases where they are used, they must contain an efficient ultra-violet screening pigment to minimise this discolouration. The most effective one is rutile titanium dioxide, but it must be used at levels of up to 50 phr to give pastel colours of limited stability. The alternative is to use high levels of inorganic pigments such as red iron oxide or green chromium oxide at 10–15 phr which will give strong dull colours in which the discolouration is less pronounced and therefore more tolerable.

Compounds of either type of colour must contain a non-discolouring type antioxidant (particularly Antioxidant 2246), which will inhibit discolouration in sunlight, and must avoid the use of aromatic oils in favour of organic ester plasticisers. These compounds are normally made from mercaptan modified types of polychloroprene which do not darken during cure, provided the curing system is free from sulphur or lead oxide.

Discolouration during vulcanisation is a specific problem in coloured cable jackets or hose covers made by the lead press method. The cause is the reaction between the lead sheath and any available sulphur in the compound, forming black lead sulphide. For this reason, sulphur modified polychloroprenes should not be used or any ingredients containing or liberating sulphur during cure, such as TMTD or factice. ETU in combination with MBTS would be the preferred accelerator.

Mercaptan modified polychloroprene types are also used when it is important to avoid the staining of an organic finish by the vulcanisate. Some sulphur modified types contain a staining antioxidant while the others may impart a slight but perceptible stain under certain conditions. Choice of antioxidant and antiozonant is of the greatest importance. It should be noted that not all non-discolouring antioxidants are non-staining to finishes. Many automotive specifications do allow slight contact staining provided there is no migration staining, enabling a preferred antioxidant such as octylated diphenylamine to be used in black compounds. When enhanced ozone resistance is needed, use can now be made of one of the recently introduced non-staining, non-discolouring antiozonants such as tris(5-norbornene 2-methyl)phosphite.

5.9. Electrical Properties

The wide use of polychloroprene in the wire and cable industry has been due to its good balance of properties, its resistance to weathering, ozone, abrasion, flame and oil. These properties make it very suitable for use in cable sheaths. Its electrical properties are rather poor in comparison with the accepted insulating materials and neoprene would be considered nowadays as an insulant only where its flame and oil resistance are indispensable.

Mercaptan modified polychloroprenes are preferred because they not only give better insulation values but are more extendable and have better extrusion properties such as speed and smoothness. However, sulphur modified (G) types are sometimes used if high tear strength is needed.

TABLE 7

HEAVY DUTY SHEATH TO BS 6899 FOR
GENERAL SERVICE

Neoprene GW	100
High activity magnesia	4
Octylated diphenylamine	2
Stearic acid	1·5
GPF carbon black	25
HAF carbon black	20
China clay	15
Antimony trioxide	3
Mineral rubber	10
Paraffin wax	4
Zinc oxide	5

Mineral fillers provide considerably higher insulation resistance and dielectric strength than does carbon black. Carbon blacks, especially the very fine particle furnace blacks (SAF and ISAF) should be avoided apart from a slight addition to give sufficient pigmentation (1–2 phr GPF black). Mineral fillers should be chosen of a type which will minimise the water absorption that will greatly reduce resistivity. Hard clays are often used while fillers such as whiting must be avoided.

Choice of plasticiser is of great importance. Most plasticisers are detrimental to the electrical properties of neoprene, but some ester plasticisers are particularly poor. These must be avoided even in jacket compounds if used without a barrier to prevent migration of plasticiser into the insulation. Naphthenic oils or hydrocarbon resins such as Kenflex A give the highest insulation values. A typical sheathing compound is shown in Table 7.

5.10. Flame Resistance

Because of the chlorine contained in the polymer, polychloroprene imparts a degree of flame resistance to vulcanisates made from it. The amount of flame resistance is dependent on compound design, especially on choice of filler and plasticiser.

Flame resistance improves with increasing levels of mineral fillers which generally provide more resistance than the equivalent volume of carbon black. Hard clays and calcium silicate give good results but the most effective filler is hydrated alumina. Care should be taken with hydrated

TABLE 8
FLAME RETARDANT CONVEYOR BELT COVER
COMPOUND

Neoprene GW	100
High *cis*-polybutadiene	4
High activity magnesia	4
Octylated diphenylamine	2
Antiozonant	2
Stearic acid	0·5
ISAF carbon black	40
Process aid	3
Paraffin wax	1
Hydrated alumina	30
Antimony trioxide	10
Flame retardant additive	10
56 % Chlorinated paraffin (liquid)	10
70 % Chlorinated paraffin (solid)	25
Zinc oxide	5
MBTS	1

alumina which, as it liberates water at high temperatures (over 180 °C), can cause porosity during high temperature curing. Highly flame resistant formulations often contain antimony trioxide and the crust-forming agent, zinc borate, in addition to hydrated alumina. Fillers such as whiting which can act as chlorine acceptors, must be avoided.

Mineral oil plasticisers which support combustion should be restricted or replaced by flame resistant plasticisers such as tritolyl phosphate or the chlorinated paraffins. These latter plasticisers, containing between 40–60 % chlorine, are preferred and often used in a blend of solid and liquid types in order to minimise the roll sticking they tend to promote (Table 8).

5.11. Aspects of Adhesion
Good building tack is found in most polychloroprene compounds but those

based on low Mooney viscosity sulphur modified (G) types are the best in this respect. When long retention of building tack is needed a crystallisation resistant copolymer (GRT) must be used. The tack may be improved by the use of coumarone–indene resins, wood resin, certain synthetic resins (e.g. Koresin) and aromatic oils. Aromatic oils are preferable since they are less likely to cause sticking of the compound to mill rolls.

Roll sticking may be eased with little loss of tack by adding three to five parts of a high cis-polybutadiene. The addition of octadecylamine (Armeen 18D) at 0·5 phr has also been found to effect release and may be used along with polybutadiene although with loss in processing safety.

Neoprene can be readily bonded to metal or textiles if certain compounding guidelines are followed. Best adhesion to metal is obtained with stocks containing carbon black and no mineral filler. Plasticisers which should be kept to a minimum, may be aromatic oils or esters, but any softener which may bloom will weaken the bond strength. For best results high moulding pressures are needed indicating the choice of a high viscosity polymer, either alone or in blend.

For adhesion to ferrous metals a two part bonding agent should be used. A primer coat is applied to the shot blasted metal followed by the tie or cover coat. For maximum adhesion, a liberal coat of high solids solution (35 % solids) based on the polychloroprene compound dissolved in toluene should be applied over the cover coat.

Brass or zinc coated steel may be bonded directly to polychloroprene compounds provided that a sulphur (1·5 phr) based curing system is used. Higher and more consistent bond strengths may be obtained by adding 5 phr of the cobalt complex Manobond C to the compound.

Adhesion to textile fibres, in a similar way, depends on the type of fibre used. The structure of cotton fibre allows a strong mechanical bond. A low viscosity polymer should be used to ensure maximum wetting and penetration of the fibre in the fabric, which must be moisture free.

Nylon and polyester fabrics need a chemical bonding agent for good adhesion. A primer coat containing 4–6 % of organic isocyanate such as Desmodur R is made by dissolving a neoprene compound in toluene to 30 % solids content. This is applied either as a dip or a spread coat to the fabric which must be stored in a moisture proof atmosphere before the final coating is applied by calender or spreading.

In the demanding application of polyester tension members in raw edge V belts, the polyester cord is pretreated by the supplier with an isocyanate priming coat followed by a resorcinol-formaldehyde/vinyl pyridine latex dip.

6.   PROCESSING DEVELOPMENTS IN POLYCHLOROPRENE

6.1. Continuous Vulcanisation of Extrusions

Four methods are currently used, apart from the specialised techniques of the cable industry, for the continuous vulcanisation of extrusions. They are the LCM bath, based on the use of an eutectic mixture of molten salts, the hot air tunnel, the fluidised ballotini bath and the microwave unit. The microwave unit, with its integral hot air tunnel has overcome many early problems including short magnetron life and uneven energy density, and has become the preferred method in Europe. It is taking over from the LCM bath which has encountered environmental problems over the disposal of the large quantities of nitrate-containing washing water. Polychloroprene is an ideal polymer for microwave curing since its polar nature enables it to absorb sufficient energy without any special compounding.

In all continuous processes vulcanisation is carried out at normal atmospheric pressure which gives rise to problems of porosity due to moisture and occluded air. The air in the compound is removed by applying 'vacuum' to the barrel of the extruder which is always of the cold feed type. The moisture in the compound is controlled by adding calcium oxide (6–10 phr) as a desiccant.

Continuously cured profiles are based on very fast curing compounds, a cure of 1 min at 200 °C being typical so that thiourea acceleration is necessary, with DETU (1 phr) most commonly used. Frequently the magnesia is reduced from the standard 4 phr to 2 phr to obtain a faster onset of cure. The processing safety of these compounds is obviously limited and requires careful control of all processing steps to limit the heat history. The compounds should be two stage mixed with the accelerators added shortly before extrusion.

In order to give the best hot green strength and freedom from distortion a high molecular weight, gel-containing mercaptan modified polychloroprene is preferred. Table 9 gives a typical formulation for a solid glazing gasket to be cured by a continuous method.

6.2. Injection Moulding[23]

The economics of injection moulding, with high capital cost of equipment and moulds, require that moulding be carried out with short high temperature cures. This is well suited for polychloroprene which, unlike natural rubber, produces optimum physical properties under these conditions. It possesses better hot tear strength than many other synthetic rubbers which is essential when moulding complicated sections such as

TABLE 9
SOLID GLAZING GASKET (CONTINUOUS CURE INCLUDING MICROWAVE AND LCM)

Neoprene TW	65
Neoprene TRT	35
High activity magnesia	2
Octylated diphenylamine	3
Factice	15
Microcrystalline wax	3
Stearic Acid	1
Petrolatum	2
FEF carbon black	25
Aromatic process oil	20
Dispersed calcium oxide	8
Zinc oxide	5
DETU	0·75

Cure: 1 min at 230 °C, 50 IRHD.

bellows and automotive boots. Neoprene GW has outstanding hot tear strength and is finding an increasing role in this type of application. Mould fouling is a more serious problem in injection moulding than the other methods of moulding. The moulds are larger, more complicated and need more time for removal and setting up after cleaning. The higher capital costs of equipment standing idle during this 'down time' make it important to minimise the problem. There is no complete answer, but a number of factors will help:

1. Use of nickel chrome steel moulds in preference to case hardened mild steel
2. Limit vulcanisation temperature to 180 °C maximum
3. Ensure adequate acid acceptor in the compound, using a high activity magnesia at up to 6 phr
4. Use the highest barrel temperature consistent with freedom from scorch
5. Minimise the use of stearates, internally and externally, as release agents

6.3. Powder Mixing[24,25]
A great deal of interest has been shown in the concept of powder mixing polychloroprene. Certain grades are commercially available in powder form, being produced from the standard chip by an attrition process and maintained in that form with the aid of 5 % of precipitated silica as

partitioning agent. Due to the high energy cost involved in this process, the powdered polychloroprene commands a surcharge. This premium is likely to limit the commercial application of powder on a large scale in the rubber industry particularly at the present time when there is excess mixing capacity available.

However, it has been shown that if the user is prepared to produce his own powdered rubber, using readily available machinery, then the surcharge can be much lower. The partitioning agent can be varied and the particle size adjusted to the optimum for a certain type of compound. The optimum particle size appears to be 1–2 mm although larger pellets up to 5 mm have been used successfully depending on the mixing unit.

The powdered polychloroprene is preblended with the compounding ingredients using a Fielder type high speed blender and then fed to the mixing equipment. This may be simply a two roll mill or an internal mixer. The use of the preblended powder compound allows shorter mixing cycles with lower energy consumption. The advantage of powdered rubber is realised only fully with one of the new continuous mixers which take the blend, mix it, vent it with 'vacuum' to remove air and moisture, and extrude it. There are two types currently available, one using the principle of an extruder with a modified screw to blend and disperse the ingredients into the polychloroprene (EVK mixer/extruder, Werner Pfliederer). The other uses a modified internal mixer to blend and disperse the compound before passing it to an extruder section (MVX powder mixer, Farrel Bridge). Both types of mixer are in commercial use.

Development work is being carried out in order to eliminate the mixer stage completely. The preblended powder is compacted and fed directly into the barrel of an injection moulding machine, where it is dispersed, blended and moulded.

BIBLIOGRAPHY

1. JOHNSON, P. R. *Rubber Chem. Technol.*, **49**, Jul.–Aug. 1976, 652.
2. MOCHEL, W. E. *J. Polym. Sci.*, **8**, 1952, 583.
3. MOCHEL, W. E. and PETERSON, J. H. *J. Am. Chem. Soc.*, **71**, 1949, 426.
4. BERCHET, G. J. and CAROTHERS, W. H. *J. Am. Chem. Soc.*, **55**, 1933, 2004.
5. WHITE, L. M., EBERS, E. S., SHRIVER, G. E. and BRECK, S. *Ind. Eng. Chem.*, **37**, 1945, 770.
6. SCHMITT, S. W. and ANOLICK, C. *Rubber Chem. Technol.*, **51**, Nov.–Dec. 1978, 888.
7. JOHNSON, P. R. *Rubber Chem. Technol.*, **49**, Jul.–Aug. 1976, 655.

8. JOHNSON, P. R. *Rubber Chem. Technol.*, **49**, Jul.–Aug. 1976, 658.
9. JOHNSON, P. R. *Rubber Chem. Technol.*, **49**, Jul.–Aug. 1976, 671.
10. SCHMITT, S. W. and ANOLICK, C. *Rubber Chem. Technol.*, **51**, Nov.–Dec. 1978, 888.
11. *Neoprene TW, TW100, TRT, Neoprene Literature*, Du Pont Publication, NP240.1, NP240TW, NP240TRT.
12. FDA. *J. Cancer Inst.*, **42**, 1969, 1101.
13. PARKES, H. G. *Rubber Industry*, Feb. 1974, 21.
14. *NA22F Neoprene Literature*, Du Pont Publication, NP 730.
15. SCHMITT, S. W. *Curing Systems for Neoprene*, Du Pont Publication, NP330.1. p. 14.
16. *Comparison Chart of the Neoprenes*, Du Pont Publication, NP210.1.
17. MURRAY, R. M. and THOMPSON, D. C. *The Neoprenes*, Du Pont Publication, p. 66.
18. MURRAY, R. M. and THOMPSON, D. C. *The Neoprenes*, Du Pont Publication, p. 78.
19. *Blend of Neoprene with Other Elastomers*, Du Pont Publication, NP380.1.
20. MURRAY, R. M. *Rubber Chem. Technol.*, **32**, Oct.–Nov. 1959, 1117.
21. SULLIVAN, R. *Compounding Neoprene to Prevent Lead Press Discolouration*, Du Pont Publication, NP580.1, p. 71.
22. SCHMITT, S. W. *Vulcanizing Methods for Neoprene*, Du Pont Publication, NP460.1, p. 8.
23. WHELANS, M. Advances in injection moulding of rubber, *Progress of Rubber Technol.*, Vol. 48, Plastics and Rubber Institute, London, 1979, pp. 195–199.
24. EVANS, C. *Powdered and Particulate Rubber Technology*, Applied Science Publishers, London, 1978.
25. PYNE, J. R. *Rubber Plastics Int.*, **3**(5 & 6), 1978, 258.
26. MURRAY, R. M. and DETENBER, J. *Rubber Chem. Technol.*, **34**, Apr.–Jun. 1961, 668.
27. MALLENBECK, A. *Compounding Neoprene for Water Resistance*, Du Pont Publication, NP520.1.

Chapter 6

BUTYL AND HALOGENATED BUTYL RUBBERS

W. D. GUNTER

Polysar Ltd, Sarnia, Ontario, Canada

SUMMARY

Butyl and halobutyl rubbers are predominant among elastomers from which, respectively, tyre inner tubes and curing bags, and tubeless tyre inner liners, are made. These are their primary applications, but various grades of the two classes of rubbers are also widely used in products as dissimilar as sealants, pipe wrapping systems, vehicle body mounts, pharmaceutical bottle closures, acid resistant tank linings, reservoir and roofing membranes, and chewing gums.

This chapter presents a brief treatment of the history, chemistry, manufacture, and applications technology of butyl and halobutyl rubbers, itemising significant developments and current trends. Most of the formulations listed do not contain ingredients that are known to be hazardous, but reference is made to lead oxides and substituted dioximes as curatives, with the expectation that compounders will give due consideration to toxicological factors when selecting compound ingredients.

1. INTRODUCTION

1.1. Distinguishing Properties of Butyl Rubbers

Standard or unmodified grades of butyl rubber are linear, amorphous, gel-free copolymers of isobutylene and isoprene. Individual grades differ from each other in functionality, the level of unsaturation varying with the ratio of isoprene to isobutylene, raw polymer viscosity (a function of molecular weight) and the properties of the stabilisers, if any, incorporated during manufacture.

The combination of properties that distinguishes butyl rubbers from highly unsaturated elastomers such as NR, SBR and NBR includes exceptionally low permeability to gases and vapours; outstanding ability to absorb mechanical energy; and excellent stability when exposed to ozone, wet and dry heat, all climatic conditions, aqueous corrosive chemicals, and animal and vegetable fats and oils.

The properties that distinguish butyl rubbers from other low-unsaturation rubbers, such as EPDM, include a higher impermeability to gases and vapours, a much greater adhesive strength, a greater effectiveness as a shock absorber over the temperature range -10 to $+80\,°C$, better retention of tear strength and extensibility on heat ageing, lower ozone resistance, lower retention of tensile strength on heat ageing, and more difficult processing.

Modified butyl rubbers—the crosslinked butyl and halogenated butyl rubbers—all possess the shock absorbency, stability and gas barrier properties that are characteristic of standard butyl rubbers. However, they differ from each other, and from standard butyl rubbers, in terms of the variables that distinguish standard butyl rubbers from each other, and in respects related to the types, amounts and functions of their modifiers. The crosslinked butyl rubbers resist flow, which is an important requirement in uncured sealant tape, their primary application. The halogenated butyl rubbers cure much faster than standard or crosslinked butyl rubbers, use less complicated curing systems and can be covulcanised with highly unsaturated elastomers.

1.2. History

Large-scale manufacture of standard butyl rubber was first achieved in the early 1940s, in the USA and Canada, shortly after those countries unexpectedly lost access to the rubber plantations in Malaysia. It was fortunate indeed that a mere five years beforehand, Sparks and Thomas[1] of the Standard Oil Co., pursuing a programme of work by I.G. Farbenindustrie on the synthesis of polyisobutylene,[2] had developed practical means of making the rubber, that essential feedstocks were available in sufficient quantities, and that the probable suitability of butyl rubber for use in tyre inner tubes had been recognised.

Annual production rose from laboratory sample quantities in 1942 to some 57×10^3 t in 1945,[3] during which time virtually all of it was consumed in inner tubes. Even today, despite the dominance of the tubeless tyre, the bulk of the standard butyl rubber produced (> 300 k t in 1979) is used in inner tubes whereas the use of natural rubber in that application is restricted, with few exceptions, to those countries in which protective tariffs discourage the use of butyl rubber.

Crosslinked butyl rubbers were first used commercially in 1967. The polymerisation technology had been developed earlier, but it had no practical value until the sealants industry identified a need for a non-curing glazing tape possessing high green strength, resilience and resistance to sag and flow.

Brominated butyl rubber was introduced to the rubber industry in the mid-1950s, following pioneering work by Morrissey and co-workers at the B.F. Goodrich Chemical Co. In a series of publications and patents issued between 1953 and 1958[4] they described the synthesis and vulcanisation of bromobutyl, and showed that halogenating butyl rubber is an effective means of increasing its cure activity without reducing its impermeability, shock absorptivity, or resistance to heat and corrosive chemicals. However, it was not until 1971, after Polysar Ltd had developed a continuous, economical manufacturing process, that a bromobutyl rubber of consistent high quality became available to the industry. Polysar Ltd was the only supplier of bromobutyl from 1971 until 1980, when Exxon began offering a competitive grade.

Commercial production of uniform, high quality, chlorinated butyl rubber was first achieved by the Exxon Chemical Co. in 1961.[5] Since then, chlorobutyl rubbers have become firmly established in the rubber industry, and market demand continues to grow. Exxon and its affiliates were the only suppliers of chlorobutyl until 1979, when Polysar Ltd entered the market with competitive grades.

Halobutyl rubbers are used in inner liners and sidewalls for tyres, heat resistant inner tubes, tank linings, pharmaceuticals, automobile mounts, and an increasing number of other applications. Annual world consumption doubled between 1974 and 1979, and is expected to exceed 100 × 10^3 t in 1980.

1.3. Manufacturing Facilities

There are now ten facilities, worldwide, producing butyl rubbers (Table 1) with an aggregate nameplate capacity exceeding 500 × 10^3 t. Esso, Exxon and the two Polysar plants include halobutyl rubbers in their product lines. Polysar Ltd is the only producer of crosslinked butyl terpolymers.

TABLE 1
PRODUCERS OF BUTYL AND HALOBUTYL RUBBERS

Facilities	Country
Cities Services Co.	USA
Esso Chemical Ltd	UK
Exxon Chemical Co. (two plants)	USA
Japan Butyl Co. Ltd	Japan
Polysar Belgium N.V.	Belgium
Polysar Ltd	Canada
Socabu	France
USSR (two plants)	Russia

1.4. Types and Grades Available

The International Institute of Synthetic Rubber Producers (IISRP) lists 35 grades of standard butyl rubbers[6] which, together with four others that are commercially available, are distributed in the unsaturation and viscosity ranges as indicated in Table 2. Staining and non-staining grades are available at most unsaturation and viscosity levels. Two grades, designed primarily for use in chewing gum and other food contact applications, do not contain a stabiliser at all.

The IISRP listings also include two crosslinked, one brominated and two chlorinated grades of butyl rubber, but three additional grades of halobutyl rubber have become commercially available since the listings were compiled.

TABLE 2
STANDARD BUTYL RUBBERS
(numbers of grades in the different unsaturation and viscosity ranges)

Unsaturation[a] (mol %)	Viscosity $(ML\text{-}1 + 8'$ at $100°C)$[a]			
	22–32	42–52	62–82	Total
0·7–0·8	0	6	4	10
1·0–1·2	4	5	0	9
1·5–1·6	0	2	3	5
1·75–2·0	0	4	6	10
2·2–2·3	0	4	1	5
Total	4	21	14	39

[a] Nominal (or mid-point of nominal range) values.

2. POLYMERISATION

2.1. Chemistry

Butyl rubber (IIR) is made by copolymerising isobutylene with small amounts (typically 0·5–3·0 parts per hundred parts of isobutylene) of isoprene (Fig. 1). The reaction occurs through an ionic mechanism, by means of which isoprene enters chains of polyisobutylene at irregular intervals (averaging approximately 200–40 isobutylene units) whose

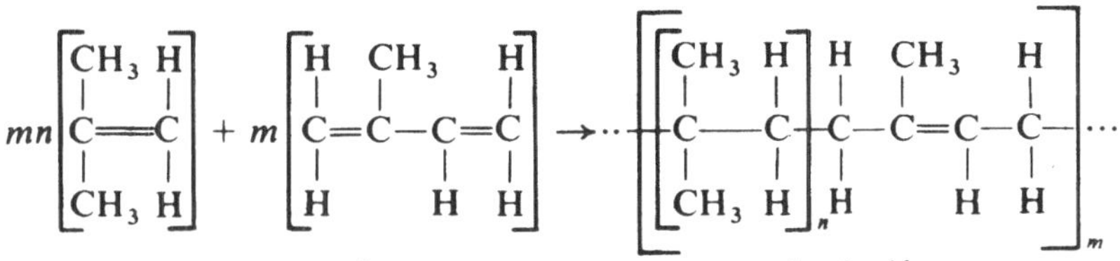

$$
mn\begin{bmatrix} CH_3 & H \\ | & | \\ C\!=\!\!=\!C \\ | & | \\ CH_3 & H \end{bmatrix} + m\begin{bmatrix} H & CH_3 & & H \\ | & | & & | \\ C\!=\!C\!-\!C\!=\!C \\ | & | & & | \\ H & H & & H \end{bmatrix} \rightarrow \cdots \begin{bmatrix} \begin{bmatrix} CH_3 & H \\ | & | \\ C\!-\!\!-\!C \\ | & | \\ CH_3 & H \end{bmatrix}_n \begin{matrix} H & CH_3 & & H \\ | & | & & | \\ C\!-\!C\!=\!C\!-\!C\!- \\ | & | & & | \\ H & & H & H \end{matrix} \end{bmatrix}_m \cdots
$$

    isobutylene   isoprene   butyl rubber

FIG. 1. Formula for butyl rubber where n averages approximately 200 (0·5 mol % unsaturation) to 40 (2·5 mol % unsaturation); and m ranges from approximately 30 to 250, depending on polymerisation conditions and mol % unsaturation, to give a weight average molecular weight typically in the range 150 000 to 550 000.

spacing depends on the ratio of the monomers in the feedstock. The resulting polymer possesses olefinic unsaturation in the range 0·5–2·5 mol %, respectively.

2.2. General Vulcanisate Properties

Vulcanisates of the polymer have excellent thermal, chemical and mechanical stability because the unsaturated isoprenyl groups are separated by long chains of saturated polyisobutylene. They are elastic because the methyl groups attached to alternate carbon atoms in the polyisobutylene chains cause steric hindrance, and this prevents crystallisation occurring readily in the unstrained state. They are highly impermeable to gases and vapours because the rate of diffusion varies with thermal mobility and the methyl groups attached to alternate carbon atoms restrict the thermal motion of the main skeletal chains. They have low resilience at temperatures in the range −40 to +75 °C because the average aggregate atomic weight of the side groups attached to each carbon atom in the chain is high (approximately 16 for IIR, 5 for NR, and 10 for SBR). Despite their low resilience, however, butyl rubbers have a low second order transition point and give vulcanisates that are flexible at temperatures down to −70 °C and below.

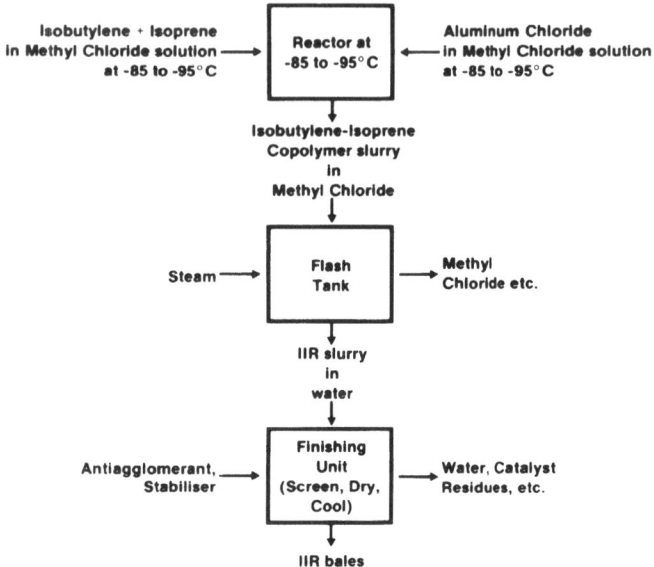

FIG. 2. Manufacture of butyl rubber.

2.3. Manufacture

In a typical industrial process, butyl rubber is made by continuously feeding methyl chloride solutions of the mixed, purified isobutylene and isoprene monomers, and of aluminium chloride catalyst, into a reactor at a carefully controlled temperature in the range -85 to $-95\,^{\circ}\text{C}$ (Fig. 2). Polymerisation is virtually instantaneous, producing a slurry of rubber particles in methyl chloride. Heat from the exothermic reaction is dissipated by continuously evaporating liquid ethylene in a cooling jacket surrounding the reactor. The slurry then passes into a flash tank where it is heated rapidly with steam. The steam flashes off the methyl chloride and other volatile residues and leaves a coarse suspension of the rubber particles in water. Sodium hydroxide is added to neutralise the catalyst residue, followed by small amounts of an antiagglomerant and, for most grades, a stabiliser whose function is to protect the rubber during the drying step. The flashed volatiles are purified and recycled, and the rubber particles are screened, dried, and compacted into bales.

The curing and processing characteristics of the butyl rubber, which are functions of its unsaturation and molecular weight respectively, are controlled by delicately balancing feedstock monomer ratio and polymerisation conditions. Aesthetic, toxicological and behavioural characteristics of the rubber are influenced by the amount and type of stabiliser, if any, incorporated during manufacture. Butyl rubbers so produced are very clean, standard analysis procedures showing, typically, less than 0·2 wt % of stabiliser, less than 0·5 wt % of volatile matter, and less than 0·5 wt % total ash. Specific gravity is 0·92.

3. COMPOUNDING OF UNMODIFIED BUTYL RUBBER

3.1. Selecting the Elastomer

When the intended application involves contact with food or drugs, then the base elastomer is selected from a small number of butyl grades. These are grades from which contaminants have been virtually eliminated and which either do not contain a stabiliser, or which contain one that is sanctioned by the appropriate regulatory authorities. For other types of application, however, selection is normally made on the basis of the level of unsaturation, the viscosity of the elastomer, and whether or not its stabiliser is staining or discolouring.

In general, as level of unsaturation increases:

Rate of cure increases	Ozone resistance decreases
State of cure increases	Weather resistance decreases
Hardness increases	Flex crack resistance decreases
Modulus increases	Elongation at break decreases
Heat resistance increases	Damping constant decreases
Resilience increases	Set decreases

Where there is a need for very rapid cures coupled with optimum resistance to heat, weather and ozone, or for maximum adhesion to highly unsaturated elastomers, the halogenated butyl rubbers should be considered. Table 3 lists examples of appropriate selections of elastomers by unsaturation level, for a variety of applications.

A grade of butyl rubber with a high raw polymer viscosity is normally needed to provide green strength when economics or technical considerations demand high filler and plasticiser loadings in the compound. Low-viscosity grades give the better calendering and extrusion properties in moderately loaded compounds. The fact that butyl rubbers do not break down significantly during mixing is an additional consideration when selecting the base elastomer.

3.2. Fillers

For practical compounds, butyl rubbers require fine-particle filler reinforcement for high tensile strength, tear strength and abrasion resistance. They also need fillers, in suitable balance with plasticisers, to enhance calendering and extrusion properties.

Carbon blacks are most effective in loadings of 30–100 parts per hundred of butyl rubber, mineral fillers are commonly used in ratios of 50–150 phr.

TABLE 3
TYPICAL ELASTOMER SELECTIONS

Standard butyl rubber		
Unsaturation (mol %)[a]		
0·7–0·8	Polysar Butyl 100 Esso Butyl 065	} Membranes, electrical insulation, mechanical goods
	Polysar Butyl 111 Esso Butyl 007	} Blends with polyolefins and waxes, chewing gum base
1·0–1·2	Esso Butyl 165	Special electricals
1·5–1·6	Polysar Butyl 301 Esso Butyl 268	} Tyre inner tubes, mechanical goods
	Polysar Butyl 101-3	Chewing gum base
1·75–2·0	Esso Butyl 365	} Heat resistant products
2·2–2·3	Polysar Butyl 402	
Crosslinked butyl rubber		
	Polysar Butyl XL-20 Polysar Butyl XL-50	} Non-curing sealant tapes and caulks
Halogenated butyl rubber		
	Polysar Bromobutyl X2 Esso Bromobutyl 2244 Esso Chlorobutyl 1066 and 1068 Polysar Chlorobutyl 1240 and 1255	} Tyre inner liners, heat resistant products, pharmaceuticals

[a] Nominal value (or mid-point of nominal) range.

Combinations of black and mineral fillers are used wherever practical to take advantage of the reinforcement achieved with the black and the processing improvements and cost reduction contributed by mineral fillers.

Whitings, talcs and soft clays enhance processing properties and are low in cost. Heat treating butyl rubbers with fine-particle precipitated silica, in the presence of a promoter, provides optimum reinforcement in light-coloured compounds provided that a small amount of diethylene or polyethylene glycol is included with the other compounding ingredients to minimise the tendency of silica to retard the cure.

Shortages of fossil fuels, and the prospect of substantial increases in carbon black prices, have recently resulted in efforts to achieve levels of reinforcement of elastomers using light-coloured fillers comparable with those attained with carbon blacks.[7] Physical and chemical means of increasing the affinity between elastomers and fine-particle white-filler surfaces are reported to be meeting with various degrees of success.

3.3. Plasticisers

Small quantities of plasticiser are used in butyl compounds in order to improve processing, reduce hardness and modulus, increase resilience and reduce cost. Mineral oils are used in most applications, paraffinic and naphthenic oils being particularly appropriate because butyl has a low solubility parameter. Waxes, tars, coumarone–indene resins and certain esters are also widely used, whereas plasticisers containing olefinic unsaturation are avoided because they retard vulcanisation.

Substantial plasticiser loadings are needed in vulcanisates that must be resilient at very low temperatures (tyre inner tubes for use in intensely cold climates, for example) and 20–30 phr, coupled with suitable filler loadings, in compounds based on high-viscosity butyl rubbers, are commonplace. Esters such as dioctyl sebacate (DOS), dioctyl adipate (DOA) and TXIB (isobutyrate ester—Eastman Chem.) are particularly suitable for low temperature performance.

3.4. Curing Systems

3.4.1. Comparison of Systems

Standard butyl rubbers are incompatible with highly unsaturated elastomers, small amounts of which can seriously interfere with vulcanisation by reacting preferentially with accelerators and curatives.

Strongly accelerated sulphur and sulphur-donor systems are the most widely used curatives for butyl rubbers because they are sufficiently adaptable to meet the technical criteria for most butyl applications at the most economical price. Activated resin systems provide outstanding resistance to hot air and superheated steam and, consequently, are used in such applications as tyre curing bags and bladders. Quinoid systems produce the fastest, tightest cures and are used primarily in room temperature curing cements and continuously cured, high voltage insulation. Peroxides cannot be used to cure unmodified butyl rubbers because they cause degradation. However, peroxides do produce a practical state of cure in brominated butyl rubbers and in crosslinked butyl terpolymers.

3.4.2. Sulphur and Sulphur-donor Systems

Less sulphur is needed to cure butyl rubbers than to cure highly unsaturated rubbers because of the relatively few sites at which crosslinking can occur. However, faster accelerator systems, higher cure temperatures and longer cure times are needed, separately or in combination. Most compounders include about 5 phr of zinc oxide in sulphur curing systems, but amounts up to 25 phr are sometimes used in compounds designed for heat resistance. Zinc oxide in excess of the amount needed for cure activation inhibits reversion by scavenging the hydrogen sulphide produced during vulcanisation, thereby enhancing heat resistance. Stearic acid is not needed for cure activation in butyl compounds but it is sometimes included as a process aid.

Combinations of thiuram and thiazole accelerators are commonly used in sulphur cures, especially where cost is a primary consideration and heat resistance is not. Dithiocarbamates give fast cures and good ageing properties, but as they tend to be scorchy a thiazole (MBT or MBTS) is normally used with them to provide processing safety. Sulphur-donor systems based on thiuram/dithiocarbamate combinations and containing very little or no elemental sulphur provide superior ageing properties and low compression set.

Typical sulphur and accelerator ranges for sulphur and sulphur-donor cure systems are shown in Table 4. Recommended combinations are shown in suppliers' literature.

The curing temperature range for butyl compounds containing sulphur-based curing systems is normally 150–185 °C but some systems will generate practical cures at much lower temperatures, given sufficient time.

3.4.3. Quinoid Systems

p-Quinone dioxime, or a suitably substituted form, used in conjunction with inorganic or organic oxidising agents will give very fast, tight cures. Quinoid systems have been used extensively in CV wire and cable insulation compounds and in adhesives that are required to cure at room temperature.

TABLE 4

Ingredient	Sulphur system (phr)	Sulphur-donor system (phr)
Thiuram and/or dithiocarbamate	1·0–2·5	3·0–4·5
Thiazole	0·5–1·0	0–0·5
Dithiodimorpholine	1·5–2·0	1·5–2·0
Sulphur	0·5–2·0	0–0·5

The crosslinks are thermally stable, high temperature ageing properties are superior to those obtained with sulphur and sulphur-donor cure systems, and ozone resistance is excellent.

Quinoid cures are, however, likely to be scorchy and to produce vulcanisates with lower than normal physical strength. Stearic acid and other acidic materials are omitted because they increase the risk of scorch. Antioxidants and antiozonants cannot be used because they react with the oxidising agent in the curing system.

Typical quinoid systems are:

p-Quinone dioxime	2 phr	2 phr
MBTS	4 phr	—
Red lead	—	6 phr

Despite the obvious merits of quinoid curing systems they have been used less extensively in recent years, partly because some of them are reported to be health hazards under some conditions, but also because of inroads made by the fastcuring halogenated butyl rubbers.

3.4.4. Resin Curing Systems[8]
Exceptional heat resistance and low compression set can be obtained by curing butyl rubbers with dimethylol phenol resins. The curing reaction is very slow even at high temperatures and when activated by halogens. Stannous chloride, and combinations of a halogenated polymer (such as neoprene, halobutyl or brominated resin) with zinc oxide, are the most commonly used halogen-bearing activators. The systems shown in Table 5 are typical.

3.5. Other Compounding Ingredients
Antioxidants and antiozonants are not needed in properly compounded butyl vulcanisates for most applications. However, for applications requiring optimum resistance to weather and ozone (e.g. electrical

TABLE 5
RESIN CURING SYSTEMS

Active phenol formaldehyde resin	10 phr	—	10 phr	7 phr
Brominated phenol formaldehyde resin	—	10 phr	—	—
Stannous chloride	2 phr	—	—	—
Neoprene	—	—	10 phr	—
Zinc oxide	—	5 phr	5 phr	5 phr
Bromobutyl	—	—	—	10 phr

insulation and reservoir membranes) it is customary to include 1·0–1·5 phr of a substituted amine, phenol or carbamate antioxidant, together with 1·0–5·0 phr of microcrystalline wax, in the compound formulation. Additional wax is beneficial for static applications because it migrates to the surface and forms a protective film.

Peptisers, too, are seldom needed in butyl rubber compounding. It is usually simpler to buy a grade of butyl rubber that already has the viscosity required, than to break down a higher-viscosity grade; the improvements in extrusion properties that peptisers produce can usually be achieved more economically with process oils. However, when a peptiser is needed, various organic peroxides, substituted aromatic mercaptans and chlorothiophenols are effective. Peroxides are very efficient peptisers for standard butyl rubbers but act as curatives for crosslinked and brominated butyl rubbers.

Promoters perform an important function in the heat treatment[9] of mixtures of butyl rubbers with fillers. Compounds based on the heat treated mixtures have better extrusion and calendering properties than they would have otherwise and their vulcanisates have greater resilience, elasticity, flexibility, abrasion resistance, and electrical resistivity. Various promoters have been withdrawn from the market recently as potential health hazards, but a dispersion of poly-dinitrosobenzene in inert filler, now marketed by Hughson Chemical as PolyDNB, is still available.

Colorants, odorants, and the many and various other materials that are incorporated in other rubber compositions can also be used with butyl rubbers, provided that they do not unduly retard or accelerate vulcanisation.

3.6. Mixing

Butyl rubber compositions are mixed, shaped and cured in conventional rubber processing equipment using techniques that are similar to those used for other types of rubber. However, particular care must be taken to prevent contamination of butyl compounds by highly unsaturated elastomers and other materials containing olefinic unsaturation, and to ensure that the final formed product does not contain unwanted, entrapped air or volatile matter. It is also important to add ingredients in a sequence that will produce good dispersion very quickly. If good dispersion is not obtained early in the mixing cycle, no amount of additional mixing later will improve it.

Processing is a branch of technology in which notable progress has been made in recent years. A more scientific approach to identifying and solving mixing problems, for example, has led to the development of 'mixing profiles'. Mixing cycles for internal mixers can be optimised in terms of scorch safety in subsequent shaping operations, energy consumption and vulcanisate properties by using such profiles.

Recent work[10] has shown that whereas upside-down mixing is the more efficient and economical method for most butyl rubber compounds, the conventional order of addition of ingredients is best for compounds that are too soft to experience the required degree of shear stress in an uninterrupted mixing cycle. Other work in the same study showed that whereas minor overloading of the mixer for a single-stage, upside-down mixing cycle for regular butyl rubbers could conceivably be beneficial under some circumstances, overloading by more than 10% is likely to be counter-productive. This degree of overloading increases power consumption, mixing time, shortens the time available to dump the batch safely and does not improve the vulcanisate properties.

In single-stage mixing it is important to keep the batch temperature low enough to avoid jeopardising scorch safety in subsequent shaping operations. When high temperature mixing is necessary, a two-stage cycle is used. The curatives are witheld until the first-stage stock has been dumped and cooled, and are then incorporated in a second stage, low temperature operation. In this way, for example, the striking improvements in processing and vulcanisate properties that can result from heat treatment are achieved without risk of scorch.

4. VULCANISATE PROPERTIES

4.1. Resistance to Heat and Long-term Ageing

Butyl vulcanisates exposed to high temperatures or extended ageing remain stable very much longer than highly unsaturated elastomers. This characteristic is attributable to their low unsaturation and to the stability of the long polyisobutylene chains between reactive sites. Yet the higher-unsaturation grades of butyl rubber give the better heat and ageing resistance.[11]

A somewhat simplistic explanation for this anomaly is that oxygen attack at reactive sites on the surfaces of rubber vulcanisates causes both crosslinking and scission of the polymer chains. The initial crosslink density and the subsequent ratio of oxidative crosslinking to oxidative chain scission are higher for vulcanisates of the more highly unsaturated butyl rubbers than for those of the less unsaturated grades. As a consequence vulcanisates of the more highly-unsaturated grades of butyl rubber retain an effective network longer. This is illustrated in Fig. 3, in which the

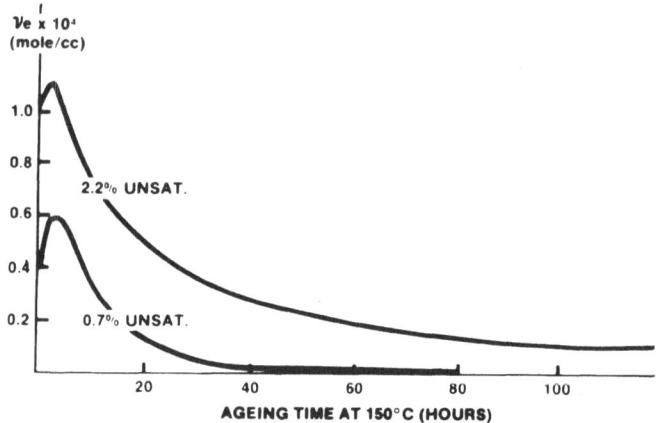

FIG. 3.   Network density versus ageing time—effect of unsaturation. Formulation of sulphur-cured gum vulcanisate (parts by weight): Rubber, 100; ZnO, 5; stearic acid, 1; MBTS, 1; TDEDC, 2; sulphur, 1·5. Cured for 20 min at 145 °C.

effective network density (derived from swelling measurements) of sulphur-cured gum vulcanisates, is shown as a function of ageing time at 150 °C. In the early stages, the network density is seen to increase, because of post-vulcanisation. Thereafter, oxidative degradation predominates and the effective network density declines, but the effective network density is clearly greater, throughout, in the more highly unsaturated grade. Since the oxidative ageing mechanism that eventually causes butyl rubber vulcanisates to fail is loss of network structure, the more highly unsaturated grades age better than the less unsaturated grades.

Sulphur curing systems provide long service in air at 100 °C or less; quinoid systems give vulcanisates with greater stability than sulphur systems and are suitable for short-term or intermittent service in air at temperatures up to 150 °C; extended service in air at 150–200 °C demands resin cures. Tyre curing bladders, for example, are virtually always made from resin cured butyl rubber. However, at high temperatures in the absence of air, e.g. in super heated, deaerated steam, properly cured butyl rubbers perform exceptionally well, irrespective of curing system.

4.2. Impermeability[12]
Among hydrocarbon elastomers, butyl rubbers are outstanding in their low permeability to gases. The process of permeation of a gas through a polymeric film involves solution of gas in the polymer at the high pressure side, diffusion through the thickness of the film, and evaporation at the low

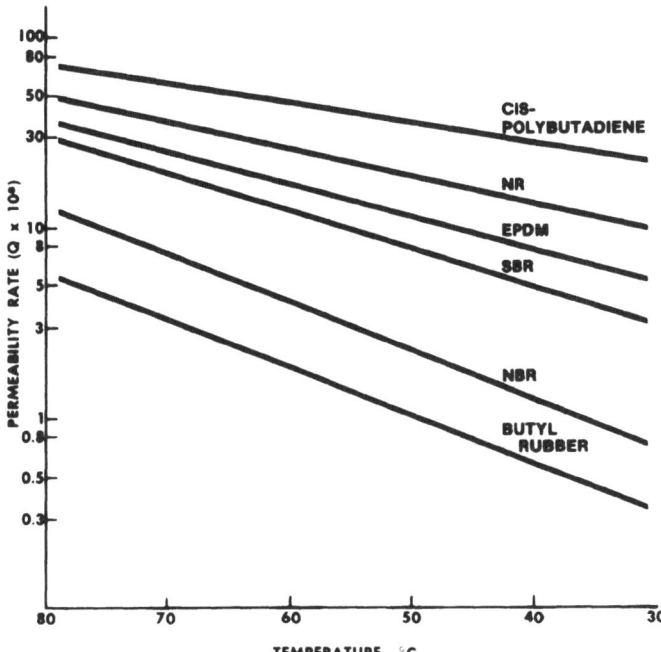

FIG. 4.   Permeability to air of vulcanisates. Q is the volume (cm^3) of air at NTP which passes (per second) through a specimen 1 cm^2 in area and 1 cm thick when the difference in pressure across the specimen is 1 atm (0·1 MPa).

pressure surface. The rate of transfer of a gas through the film is governed by the solubility of the gas (which defines the concentration gradient within the specimen) and the diffusivity (which defines the rate of migration of the gas molecules under this concentration gradient).

The solubility of gases in butyl rubber is similar to their solubility in other hydrocarbon elastomers. However, rate of diffusion varies with thermal motion, and the methyl groups attached to the main-chain backbone restrict thermal motion, so that the rate of diffusion of gases through butyl rubber is exceptionally low. This phenomenon is one of the most important intrinsic properties of butyl rubber and is responsible for many of its major uses; tyre inner tubes being the prime example. However, it is also a source of difficulty during processing because pockets of entrapped gases will not dissolve in, and diffuse readily into, the surrounding rubber.

Figure 4 shows how vulcanisates of several common elastomers compare in terms of permeability to air. The vulcanisates contained 100 parts of polymer, 5 phr of zinc oxide, 1 phr of stearic acid, 50 phr of carbon black, and a curing system appropriate to each particular polymer. Although differences in permeability decrease with increasing temperature, it can be seen that the natural rubber vulcanisate is still at least 10 times as permeable as the butyl rubber vulcanisate at the temperature of a running tyre.

The importance of compounding variables is secondary to polymer type where permeability is concerned, but, in general, permeability decreases with increased filler loading and increases with increased plasticiser loading.

4.3. Ozone Resistance

Butyl rubbers are capable of a high level of ozone resistance. Compounding principles which favour ozone resistance include low unsaturation, high state of cure, avoidance of excessive loadings of plasticisers and coarse non-black fillers, thorough dispersion and wetting of compounding ingredients, and a judicious choice of antiozonants and waxes in those compounds in which they are needed. Conformity with these principles provides complete protection against atmospheric ozone in static applications and a much higher level of ozone resistance than is achievable with highly unsaturated, general purpose elastomers in dynamic applications.

4.4. Weathering

The weathering of black butyl rubber vulcanisates is outstanding. In light-coloured goods, excellent weathering is achieved using highly reflective pigments such as zinc oxide and titanium dioxide. High loadings of highly reinforcing white fillers can be detrimental because they cause surface embrittlement, and consequent surface crazing, at relatively low levels of polymer degradation.

4.5. Resilience and Low Temperature Properties

Cured butyl rubber possesses exceptional ability to absorb vibration and impact energy at temperatures up to 100 °C, yet it remains flexible down to −73 °C (its glass transition point). This combination of properties is illustrated in Fig. 5 which plots ball rebound against temperature for vulcanisates of a number of common elastomers. Butyl rubber is distinctive in the very broad range of temperatures above and below room temperature at which rebound values are very low. The high deformation rate in this test is characteristic of the rates encountered in vibrating automotive parts, engine mounts and running tyres. Low rebound represents a large capacity to absorb mechanical energy under these conditions, thus damping out undesirable vibrations.

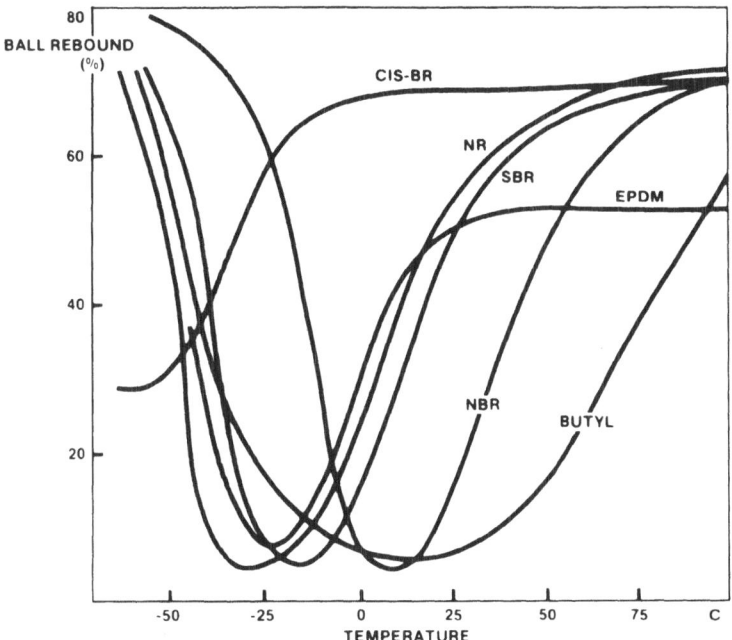

FIG. 5.   Ball rebound versus temperature.

The damping properties of butyl rubber at high deformation rates are inherent in the polyisobutylene structure and are not greatly affected, in the usual service temperature range, by level of unsaturation, state of cure or compounding variations. Butyl rubber remains flexible at temperatures down to its glass transition temperature but its rate of recovery from deformation at low temperatures is very slow. Optimum low temperature retraction rate in butyl rubber compounds is achieved using a high level (25 phr or more) of a light hydrocarbon or ester plasticiser to provide ample internal lubrication, together with a high molecular weight, highly unsaturated butyl polymer, heat treatment, and a tight cure.

4.6. Water Absorption

Butyl rubber is capable of very much lower water absorption rates than other common elastomers, partly because its structure restricts thermal motion (which is essential for diffusion), but mainly because it contains virtually no electrolytic impurities. This latter reason is, of course, a factor in the selection of butyl rubber for electrical insulation.

If water-soluble substances are introduced when compounding the rubber, moisture will eventually diffuse to the water-soluble particles and form solution droplets within the rubber matrix. The difference in the vapour pressure of the internal solution and the vapour pressure of any external water phase then becomes the driving force for further infusion of water. When the rubber matrix is vulcanised, the elastic modulus of the rubber network exerts pressure on the internal solution, and when this pressure equals the osmotic pressure of the internal solution, equilibrium is established and absorption ceases.[13] Consequently, when compounding for minimum water absorption it is important to minimise the introduction of electrolytic substances into the compound and to select the grade of butyl rubber and curing system which will give the highest elastic modulus consistent with other performance needs.

4.7. Chemical Resistance
Butyl rubber vulcanisates are exceptionally resistant to animal and vegetable fats and oils. They are also highly resistant to dilute and moderately concentrated oxidising acids, non-oxidising acids, aqueous bases, oxidising and reducing solutions, and polar solvents. However, they are not serviceable in hydrocarbon oils and solvents, which cause gross swelling, or in concentrated oxidising acids.

The more highly unsaturated butyl rubbers, tightly cured, containing low levels of highly reinforcing fillers, and heat treated, give best performance in terms of vegetable oil, chemical, and polar solvent resistance. Fillers and other compounding ingredients are chosen for their inertness when in contact with the chemicals to be resisted and should be thoroughly dispersed so that they are wetted by a continuous polymeric phase.

4.8. Abrasion Resistance
Butyl rubber vulcanisates are characteristically strong, tough and tear resistant, and have ample abrasion resistance for applications in which butyl rubbers are normally used. Highly reinforcing carbon blacks and heat treatment contribute to optimum results.

5.   APPLICATIONS OF UNMODIFIED BUTYL

5.1. Tyre Inner Tubes
Since butyl rubbers were first manufactured, inner tubes have been their most important application, despite the advent of tubeless tyres. This

TABLE 6
BIAS-PLY TYRE INNER TUBE

Formulation (parts by weight)	
Polysar Butyl 301[a]	100
Stearic acid	1
Zinc oxide	5
N-660 (GPF) black	70
Paraffinic oil	27·5
MBT	0·5
TMTD	1
Sulphur	1·5
Compound viscosity (ML-4' at 100 °C)	45
Mooney scorch (t_5 at 125 °C) (min)	>25
Garvey Die extrusion, No. 1 Royle, 104 °C, 30 rpm	
Rate (cm/min)	64
Die swell (%)	60
Appearance	B10
Cured 10 min at 171°C	
Hardness, Shore A	48
Modulus at 100% elongation (MPa)	1·4
Modulus at 300% elongation (MPa)	4·8
Tensile strength (MPa)	11·9
Elongation at break (%)	640
Tear strength, Die C (kN/m)	38
Tension set, Smithers (%)	16
Compression set, ASTM B (%)	
22 h at 125 °C	82
Aged in air, 70 h at 125°C	
Hardness, Shore A	55
Tensile strength (MPa)	9·5
Elongation at break (%)	460

[a] IIR, 1·6 mol % unsaturation, 55ML-4' at 125 °C, Polysar Ltd.

application exploits the exceptional impermeability, heat resistance, durability, tear resistance and low temperature flexibility of butyl rubbers.

Splicing, valve bonding and puncture repairing techniques were developed quickly, and a 'cold buckling' problem[14] (caused by the slow shape retraction of inner tubes at temperatures below -20 °C and manifested by circumferential folding of the tube inside the tyre, chafing of the tube wall, and consequent tube failure) was solved by the manufacturers of butyl rubbers in the late 1940s. These events established industry-wide inner tube compounding and manufacturing practices which changed very

TABLE 7
RADIAL-PLY TYRE INNER TUBE

Formulation (parts by weight)	
Polysar Butyl 301[a]	100
PolyDNB[b]	0·2
Stearic acid	1·5
Zinc oxide	5
N-550 (FEF) black	47·5
Naphthenic oil	20
TMTD	1
Sulphur	1·75
Compound viscosity (ML-4′ at 100 °C)	47
Mooney scorch (t_5 at 125 °C) (min)	25
Garvey Die extrusion, No. 1 Royle, 104 °C, 30 rpm	
Rate (cm/min)	180
Die swell (%)	204
Appearance	AB10
Cured 10 min at 165 °C	
Hardness, Shore A	46
Modulus at 100 % elongation (MPa)	1·0
Modulus at 300 % elongation (MPa)	3·9
Tensile strength (MPa)	14·2
Elongation at break (%)	750
Tear strength, ASTM Die C (kN/m)	40
Tension set, Smithers (%)	26
Compression set, ASTM B (%)	
22 h at 125 °C	82
Aged in air, 70 h at 120 °C	
Hardness, Shore A	51
Tensile strength (MPa)	10·9
Elongation at break (%)	580

[a] IIR, 1·6 mol % unsaturation, 55 ML-4′ at 125 °C, Polysar Ltd.
[b] Promoter, 30 % poly-*p*-dinitrosobenzene + inert filler, Hughson.

little over the years, until the emergence of radial tyres and, later, halogenated butyl rubbers.

Typical butyl tube compounds for bias-ply and radial-ply tyres are shown in Tables 6 and 7. (Halobutyl tube compounds are described later.)

The first is an easy processing, low cost compound. The grade of rubber selected cures faster than low-unsaturation grades, accommodates the high oil loading needed to prevent cold buckling and provides excellent ageing and air retention.

The second compound contains a smaller carbon black loading because a

TABLE 8
TYRE CURING BAG/BLADDER

Formulation (parts by weight)	
Polysar Butyl 301[a]	95
Neoprene W[b]	5
Stearic acid	1·5
Zinc oxide	5
N-347 (HAF-HS) black	50
Aromatic oil	7·5
SP-1045[c]	7
Compound viscosity (ML-4' at 100 °C)	75
Mooney scorch (t_5 at 125 °C) (min)	25
Cured 60 min at 165°C	
Hardness, Shore A	62
Modulus at 100 % elongation (MPa)	1·7
Modulus at 300 % elongation (MPa)	7·1
Tensile strength (MPa)	14·8
Elongation at break (%)	580
Tear strength, Die B (kN/m)	58
Tested at 135°C	
Tear strength, Die C (kN/m)	39
Modulus at 100 % elongation (MPa)	1·6
Modulus at 300 % elongation (MPa)	5·1
Tensile strength (MPa)	7·5
Elongation at break (%)	460
Tear strength, Die B (kN/m)	26
Tear strength, Die C (kN/m)	18
Aged in air, 70 h at 177°C–change	
Hardness (points)	+11
Modulus at 100 % elongation (%)	+76
Tensile strength (%)	−60
Elongation at break (%)	−60
Aged in saturated steam, 24 h at 1·55 MPa—change	
Hardness (points)	+2
Modulus at 100 % elongation (%)	+28
Tensile strength (%)	+3
Elongation at break (%)	−24
Volume change (%)	+2

[a] IIR, 1·6 mol % unsaturation, 55 ML-4' at 125 °C, Polysar Ltd.
[b] CR, non-sulphur-modified, Du Pont.
[c] Curative, active phenolic resin, Schenectady.

low modulus is needed to accommodate the greater flexing and chafing to which inner tubes are subjected in radial-ply tyres. FEF black was selected in this instance for ease of extrusion and resistance to abrasion between tube and tyre, but N-347 (HAF-HS) black is also widely used. Heat treatment of the rubber, promoter, black and oil for at least 2 min at 165 °C is essential for good extrusion when the black level is so low.

Black and oil loadings reflecting the differences in properties required in tube compounds for bias-ply and radial-ply tyres are shown in Table 19 (see Section 7.6.3).

5.2. Tyre Curing Bag/Bladder
These products are almost exclusively made from butyl rubbers. A typical formulation is shown in Table 8. Resin curing provides excellent resistance to wet and dry heat. Polychloroprene supplies the halogen needed to activate the resin. High structure black provides processing ease and high thermal conductivity.

5.3. Waterproofing Membrane
A typical formulation is shown in Table 9. Butyl rubber is used because of its moisture impermeability, bond strength and flexibility over the full service temperature range. The low-unsaturation grade is selected for optimum resistance to weather and ozone. The filler–plasticiser combination and low raw polymer viscosity provide the low compound viscosity needed for good calendering properties. Furnace black enhances abrasion resistance. Substituting 20 parts of EPDM for 20 parts of the butyl rubber would produce even better weathering and ozone resistance but at some cost in splice strength and extensibility on ageing.

5.4. Sewer Gasket
Butyl rubber has very low water absorptivity and is highly resistant to bacteria, animal and vegetable matter, and the chemicals used to clean and clear sanitary drainage and disposal systems. The high-unsaturation grade provides fast, tight cures. The rubber, promoter, carbon black and oil in the compound shown in Table 10 need to be heat treated at 165 °C for at least 2 min, and the vulcanisate needs to be post-cured, for low compression set.

5.5. Solvent Release Sealants
This is a large and growing application for butyl rubbers. Butyl caulks are used in limited movement lap and butt joints, and in concealed or interlocking metal-to-metal joints such as are found in curtain wall

TABLE 9

WATERPROOFING MEMBRANE

Formulation (parts by weight)	
Polysar Butyl 100[a]	100
Stearic acid	1
Zinc oxide	5
N-330 (HAF) black	50
N-550 (FEF) black	20
NBC	1·5
Paraffin wax	2
Soft clay	10
Petrolatum	5
TMTD	1
MBT	1
CuMDC	0·5
MC sulphur	1
Compound viscosity (ML-4′ at 100°C)	72
Mooney scorch (t_5 at 125°C) (min)	20
Cured 50 min in dry air at 153°C, 0·35 MPa	
Modulus at 300% elongation (MPa)	7·9
Tensile strength (MPa)	13·7
Elongation at break (%)	510
Tear strength (kN/m)	53
Brittle point (°C)	−41
Static ozone resistance	
(50 pphm O_3, 168 h, 40°C, 20% extension)	no cracks
Aged in air, 168 h at 121°C—change	
Tensile strength (%)	−6
Elongation at break (%)	−18

[a] IIR, 0·7 mol % unsaturation, 51 ML-8′ at 100°C, Polysar Ltd.

construction, in factory assembled building components, in heating and air conditioning ducts, and in car and trailer bodies.

Low-viscosity grades of butyl rubber are normally used because IIR has a high solution viscosity, and low-unsaturation grades give optimum weathering. The compound described in Table 11 gives a good quality caulk at relatively low cost and meets US Interim Specification TT-S-001657. Asbestos has been eliminated from the formulation by using a combined methanol activated and heat activated thixotrope system.

5.6. Butyl/Wax Blends

Butyl/wax coatings are used to improve the barrier properties of paper and

TABLE 10

SEWER GASKET

Formulation (parts by weight)	
Polysar butyl 402[a]	100
Promoter	0·2
Zinc oxide	5
Stearic acid	1
N-990 (MT) black	45
N-550 (FEF) black	10
Paraffinic oil	20
DTDM	2
TMTD	3
Compound viscosity (ML-4' at 100 °C)	36
Mooney scorch (t_5 at 125 °C) (min)	25
Cured 20 min at 165 °C[b]	
Hardness, Shore A	53
Modulus at 300 % elongation (MPa)	6·9
Tensile strength (MPa)	9·7
Elongation at break (%)	430
Compression set, ASTM B (%)	
22 h at 70 °C	9
70 h at 120 °C	15
Aged in air, 70 h at 100 °C—change	
Hardness (points)	−1
Modulus at 300 % elongation (%)	+12
Tensile strength (%)	+5
Elongation at break (%)	−10

[a] IIR, 2·2 mol % unsaturation, 45 ML-8' at 100 °C, Polysar Ltd.
[b] Post-cured 2 h in air at 130 °C.

cellophane food wrappings and containers. The butyl rubber increases the flexibility of the coatings at low temperatures, minimises fold cracking, increases impermeability to moisture, increases resistance to vegetable oils at high temperatures, increases scuff resistance, and reduces the tendency of waxes to melt and flow at moderately elevated temperatures.

Blends containing 5 % or more of a low-unsaturation, food-grade butyl rubber (such as Polysar Butyl 111), together with 0·25 % by weight of a suitable antioxidant, can be maintained at 120 °C for at least 30 h without loss in melt viscosity. This is important in paper waxing processes in which coating efficiency depends upon a carefully sustained balance of roll speed, roll pressure and melt viscosity.

TABLE 11
SOLVENT RELEASE SEALANT (CAULK)

Formulation (parts by weight)
Polysar Butyl 100[a]	100
Mineral spirits	100
Antioxidant	1
Calcium carbonate	300
Talc	100
Titanium dioxide (rutile/anatase)	10
Bentone 38[b]	10
Methanol/water (95:5)	5
Thixatrol GST[c]	10
Polyterpene resin tackifier	12
Polybutene plasticiser (low MW)	70
Paraffinic oil	12
Mineral spirits	15
Bubble formation (%)	18
Tenacity	no cracking separation or adhesion loss
Shrinkage (%)	20·7
Slump (mm)	2·5
Extrudability (s/ml)	4·3
Stain index	1·0
Tack-free time (adhesion to polyethylene after 24 h)	none
Aged property rating	
Edge cracking	none
Centre cracking	none
Adhesion loss	none
Colour change	white to ivory
Bond cohesion loss (cm²)	
Aluminium	0
Glass	0
Mortar	0

[a] IIR, 0·7 mol% unsaturation, 51 ML-8′ at 100°C, Polysar Ltd.
[b] Chemically activated clay thixotrope, N.L. Industries.
[c] Heat activated castor oil-based thixotrope, N.L. Industries.

5.7. Butyl/Polyolefin Blends

Modification of polyolefins with an appropriate grade of butyl rubber produces the following important improvements in their properties:

Increased resistance to environmental stress-cracking, impact strength, tear strength, impermeability to gases and vapours, flexibility, and filler loading capacity.

Reduced incidence of voids when thermoforming.

Such improvements are of value in a variety of polyolefin applications, including:

Aerosol tubing
Films/sheeting
Wire and cable insulation/moulded electricals
Containers for industrial chemicals (oxidising acids, alcohols, etc.),
    foods/dairy products, and soaps/wetting agents
Appliances/auto parts/housewares
Bottle caps/closures for chemicals/pharmaceuticals

The effects vary with the type of polyolefin and the blend ratio; butyl rubber is more effective than other modifiers in some respects and at some ratios.[15]

6.   CROSSLINKED BUTYL RUBBERS

6.1. Terpolymer Systems

Crosslinked butyl rubbers were first used commercially in 1967, when the sealants industry identified a need for a non-curing, butyl-based sealant tape having high green strength, resiliency (recovery after deformation) and a high level of resistance to sag and flow. These properties were achieved by copolymerising isobutylene and isoprene with divinylbenzene so as to form crosslinked terpolymers. The extent of the crosslinking (and thus the levels of sag and flow resistance, resiliency and green strength) are controlled by the level of divinylbenzene in the polymerisation feed stream. The terpolymer is used in the uncured state in most of its applications, but it can be cured with any of the systems used to cure unmodified butyl rubbers.

The manufacturing process for butyl terpolymers involves only minor changes to the procedure and equipment used for unmodified butyl rubbers, and, being continuous, generates products with a high level of uniformity.

The two commercially available grades of crosslinked butyl rubber (Polysar Butyl XL-20 and XL-50) have solubilities in cyclohexane of 20 % and 50 %, respectively. The former, the more highly crosslinked type, is used primarily as the elastomeric base for the premium quality, preformed, non-curing tapes used to seal, retain and cushion the windscreens and other fixed windows of vehicles. The latter type too is used mainly in the uncured state, and is highly suitable for use in preformed automotive glazing sealants for which the greater green strength of the more highly crosslinked type is either unnecessary or inappropriate. It is also used in some compounds instead of a promoter, as a processing aid and to provide green strength. However, its main applications are in preformed sealant tapes for curtain-wall construction, in hot-melt sealants, and in high performance, solvent release sealants (caulks).

Several other compositions that are commonly classed as crosslinked butyl rubbers are available, although they are not listed as such by their manufacturers, in *The Synthetic Rubber Manual—8*.[6] They are made by partially curing compounds of unmodified butyl rubber containing additional materials such as fillers, plasticisers and resins; they too are used primarily in caulks and sealants. They do not offer the high levels of resistance to sag and flow that are achievable with the terpolymers, but they are much easier to process.

TABLE 12
WINDSCREEN SEALANT TAPE

Formulation (parts by weight)	
Polysar Butyl XL-20[a]	100
Polybutene	160
Terpene phenol resin	20
N-550 (FEF) black, beads	140
Solids, 3 h at 102 °C (%)	99·4
Ash, 1 h at 760 °C (%)	15·1
Specific gravity	1·08
Hardness, Shore A	
Unaged	12
Aged 336 h at 88 °C	17
Aged 500 h at 37 °C, 95 % RH	15
Aged 500 h in UV light	16
Migratory staining	none
Paint incompatibility	
(stains, pinholes, blistering, softening)	none
Low temperature flex cracking	none
Heat damage	
(flow, blistering, adhesion loss)	none
Rebound height (mm)	5·9
Flow, 336 h at 88 °C (mm)	0·76
Height, after 24 h flow (mm)	5·8
Force to compress (N)	409
Yield strength (kPa)	46
Cohesive failure rate (%)	100
Self stick	good
Fatigue resistance (kc)	> 250

[a] Crosslinked IIR, 20 wt % solubility in cyclohexane.

TABLE 13
HIGH QUALITY CAULK

Formulation (parts by weight)	
Polysar Butyl XL-50[a]	100
Antioxidant	1
Calcium carbonate	620
Precipitated hydrated silica	40
Polybutene plasticiser (high MW)	190
Thixatrol ST[b]	30
Synthetic polyterpene tackifier	20
Varsol 3139[c]	159
Tack-free time	no transfer
Extrudability (g)	390
Weight loss after heat ageing (%)	20
Sag and flow	
Vertical (mm)	1·6
Horizontal (mm)	0
Hardness, Shore A	<40
Bleeding (mm)	<3
Tensile extension (after 96 h in	
UV light, then 96 h in water)—	
force to 20% extension from substrate	
Mortar (N)	23
Aluminium (N)	30
Glass (N)	32
Bond area loss (cm²)	<2·0
Peel adhesion, 180°	
Mortar (kN/m)	0·7
Aluminium (kN/m)	0·7
Glass (kN/m)	1·7
After exposure to UV light (kN/m)	1·5
Low temperature flexibility	no loss of adhesion
Cracks and blisters	none

[a] Crosslinked butyl, 50% soluble in cyclohexane, Polysar Ltd.
[b] Heat activated castor oil thixotrope, N.L. Industries.
[c] Aliphatic solvent, Esso.

High-shear mixing equipment is needed to mix the terpolymer with the higher level of crosslinking (XL-20). The other terpolymer (XL-50, 50% soluble in cyclohexane) is considerably easier to mix, but it too requires medium- to high-shear mixing equipment for satisfactory dispersion. Upside-down mixing techniques have been developed to alleviate the problem but some degree of mixing difficulty appears to be the price that has to be paid for optimum green strength, resiliency and resistance to sag and flow.

6.2. Windscreen Sealant Tape[16]

The compound shown in Table 12 meets specification ESB-M3G95-AB of the Ford Motor Co.

6.3. High Quality Caulk (Solvent Release Sealant)

Crosslinked butyl rubbers provide excellent bases for high quality butyl caulks. Their inherent strength and resistance to flow allows considerable latitude in compound design. Sealants based on these rubbers possess higher strength, extensibility, recovery, resiliency and a greater resistance to slump than comparable sealants based on regular butyls, resulting in products with properties closer to those found in the more expensive cured liquid sealants. Crosslinked butyls are capable of accepting the extra high loadings of fillers and plasticisers required to optimise the balance of physical and rheological properties.[17]

The compound described in Table 13 combines high quality with relatively low cost and meets Canadian specification CGSB 19-GP-14M.

7. HALOGENATED BUTYL RUBBERS

7.1. Chemistry

Halogenated butyl rubbers are made by reacting elemental bromine or chlorine with butyl rubber dissolved in a light (C_5-C_8) aliphatic hydrocarbon such as hexane. The olefinic unsaturation remains, but the double bonds move from the backbone to adjacent pendant positions, and the halogen attaches itself to the carbon atom which is allylic to the double bonds (Fig. 6). Commercial halobutyls contain, typically, $1\cdot1-1\cdot3$ wt % of chlorine or $1\cdot9-2\cdot1$ wt % of bromine.

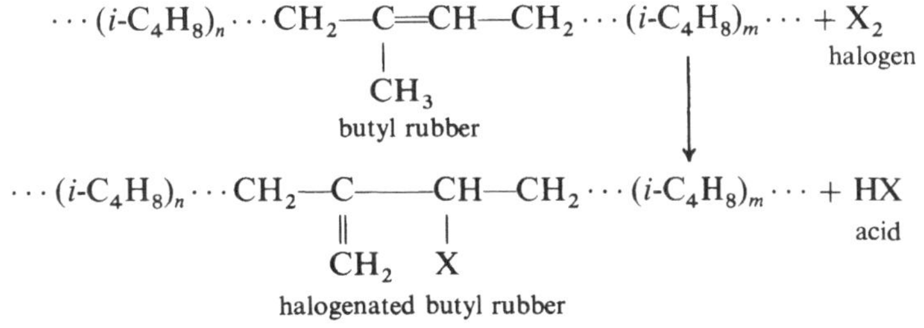

$$\cdots(i\text{-}C_4H_8)_n\cdots CH_2-\overset{\displaystyle |}{\underset{\displaystyle \underset{\text{butyl rubber}}{CH_3}}{C}}=CH-CH_2\cdots(i\text{-}C_4H_8)_m\cdots + X_2 \quad \text{halogen}$$

$$\cdots(i\text{-}C_4H_8)_n\cdots CH_2-\overset{\displaystyle \|}{\underset{\displaystyle CH_2}{C}}\underset{\displaystyle \underset{\text{halogenated butyl rubber}}{X}}{\overset{\displaystyle |}{CH}}-CH_2\cdots(i\text{-}C_4H_8)_m\cdots + HX \quad \text{acid}$$

FIG. 6. Formation of halogenated butyl rubber.

The presence and location of the halogen increases the reactivity of the olefinic double bonds and supplies additional, highly reactive sites for crosslinking. The polyisobutylene chain length is unchanged, provided that the feed ratio of halogen to bound isoprene is not excessive, and the bulk and steric hindrance associated with the pendant groups remain. Consequently, halogenated butyl rubbers retain most of the distinctive and desirable properties of unmodified butyl rubbers, and offer important advantages related to their cure versatility and fast cure rate. They can be cured with zinc oxide alone as they cure very much faster and reach a higher state of cure than unmodified butyl rubbers. They can be covulcanised with highly unsaturated rubbers or they can be cure-bonded to them; this characteristic is vital to their use in tyre inner liners. Halobutyl vulcanisates are comparable with butyl vulcanisates in their stability in air at high temperatures and they have lower set. They retain extremely low permeability to gases, high hysteresis, and excellent resistance to weather, ozone, flex cracking, animal and vegetable fats and oils, and corrosive chemicals.

Halobutyl rubbers cannot be compounded to give the extraordinary level of moisture resistance that may be achieved with butyl vulcanisates because some of their vulcanisation by-products are water-soluble salts. For this reason they are not used in high-voltage electrical insulation but they possess sufficient moisture resistance to give outstanding performance in hose tubes and covers for hot water and superheated steam.

7.2. Manufacture

Elemental chlorine or bromine is brought into intimate contact with butyl rubber in a hydrocarbon solvent solution. The hydrochloric or hydrobromic acid by-product is neutralised with dilute caustic soda, and the halogenated butyl rubber is recovered from the hydrocarbon solution by conventional steam stripping. The resulting slurry of halobutyl in water is screened, dried and compacted into bales. Halogenation is rapid and the commercial process is continuous.

7.3. Compounding
7.3.1. Selecting the Elastomer
Bromobutyl rubbers are normally selected when optimum performance is required, for example, in inner liners and inner tubes for heavy-duty and premium quality tyres. Chlorobutyl rubbers are selected when service demands are less severe but exceed the capabilities of non-halogenated butyl rubbers, for example, in covulcanisable blends with natural rubber

for tyre inner liners and white sidewalls. Standard butyl rubbers cost less than the halobutyl rubbers and are selected when their properties are fully adequate, for example, in inner tubes for passenger vehicle tyres. There are important applications in which non-halogenated butyl rubbers perform better than halogenated types, despite being lower in cost. For example (i) in tyre curing bladders and electrical insulation—in which their lower moisture absorption is a distinct advantage, and (ii) in non-curing sealants (tapes and caulks)—here 'structure' of the crosslinked butyl grades gives them an advantage.

Bromobutyl rubbers are more expensive to produce, and command a higher price, than chlorobutyl rubbers but they provide higher green strength, cure faster, require less curative for an equivalent state of cure, provide stronger cured adhesion to highly unsaturated rubbers, and provide better flex crack resistance both before and after heat ageing. They can also be cured with organic peroxides, whereas the chlorobutyl rubbers cannot, and can therefore be considered for use in applications in which the presence of zinc or sulphur in their compounds would be objectionable.

The differences in the cure reactivity of chlorobutyl and bromobutyl rubbers are sufficiently great so that one cannot usually be substituted directly into compounds designed specifically for the other. Bromobutyl substituted for chlorobutyl in a chlorobutyl compound would be likely to prove unacceptable for factory processing because of a greatly increased risk of scorch. Chlorobutyl substituted for bromobutyl in a bromobutyl compound would be unlikely to reach an adequate state of cure in a reasonable time. In other respects, however, the compounding principles and practices used for unmodified butyl rubbers apply also to the halogenated butyl rubbers.

7.3.2. Curing Systems
7.3.2.1. General. The cure reactivity associated with the presence of halogen in the polymer molecule is inversely related to the strength of the bond by which the halogen is attached. Consequently, chlorobutyl and bromobutyl rubbers have quite different cure reactivities. The difference is such that at one extreme organic peroxides, or even sulphur alone, will cure bromobutyl rubber but will not affect the less reactive chlorobutyl rubber, and at the other extreme, certain amines provide safe, practical cure systems for chlorobutyl rubber, but immediately scorch bromobutyl rubber.

Between those extremes, however, both bromobutyl and chlorobutyl rubbers have considerable cure versatility. Both produce excellent vulcanisates when cured with 3 phr of zinc oxide alone, which is important

TABLE 14

CURE SYSTEM GUIDELINES—100% BROMINATED BUTYL

Cure system	Scorch safety	Heat resistance	Compression set	Flex	Adhesion
Sulphur	Fair	Poor	Fair	Fair	Fair
Zinc oxide	Very good	Good	Fair	Good	Fair
ZnO + TCBQ	Very good	Excellent	Fair	Fair	Fair
ZnO + TMTD + MgO	Good	Excellent	Good	Good	Very good
ZnO + MBTS + TMTD + MgO	Good	Very good	Fair	Very good	Excellent
ZnO + ZDC	Poor	Good	Very good	Fair	Very good
Phenolic + ZnO	Good	Excellent	Excellent	Fair	Fair
Peroxide + co-agent	Very good	Excellent	Excellent	Fair	Fair

TCBQ = Tetrachlorobenzoquinone.

in some pharmaceutical applications, but in industrial practice most systems contain sulphur, sulphur-donors, accelerators, or phenolic resins, in two-, three-, or four-component combinations with zinc oxide.

The role of magnesium oxide in halobutyl curing systems is variable. It is commonly used in amounts ranging from 0·1 to 0·5 phr. In chlorobutyl compounds, used in amounts up to 0·2 phr, it increases scorch safety without significantly affecting stress/strain properties. When used in higher amounts with some curing systems it substantially slows the cure rate and depresses stress/strain properties. It has the reverse effect, however, if used in chlorobutyl compounds with amine cure systems. In bromobutyl compounds its role is obscure, but it does contribute to processing safety, shelf life and heat resistance when used with curing systems that do not contain elemental sulphur.

An impression of the combinations of properties achievable with various bromobutyl cure systems can be gained from the data in Table 14.

7.3.2.2. Sulphur systems. Conventional cure systems for most current, practical, tyre-related and mechanical goods formulations consist of zinc oxide, plus sulphur or sulphur-donors, accelerated with sulphenamides or benzothiazoles. These types of accelerator function initially as retarders but ultimately produce very high states of cure. MBT (mercaptobenzothiazole) is an exception in that it will scorch bromobutyl stocks, but it can be used safely in some light-coloured, chlorobutyl-based compounds which need the stronger acceleration that MBT can provide.

7.3.2.3. Resin curing systems. Phenolic resins used in small amounts in conjunction with zinc oxide, cure halogenated butyl rubbers quickly and give vulcanisates that have high modulus and excellent dry heat resistance. The small amount of resin needed, and the rate of cure achieved, provide a double advantage over resin curing regular butyl rubbers but the resistance of the vulcanisates to steam is less than that of resin cured regular butyl. Additional halide is not needed in bromobutyl compounds to activate the resin because of the activity of the bromine that is already present, but brominated dimethylol phenol resins may be needed in chlorobutyl compounds for fast, tight cures.

7.3.2.4. Peroxide systems. Peroxide systems are sometimes used to cure bromobutyl compounds[18] for use in pharmaceutical products, in which normal vulcanisation by-products would be objectionable, and in applications requiring extremely low compression set. They are seldom used otherwise. Peroxides will not cure chlorobutyl rubbers, and an adequate state of cure in bromobutyl compounds can normally be achieved only when a coagent such as HVA No. 2 (N,N',*m*-phenylenedimaleimide, Du Pont) is present. Stocks containing peroxide/coagent cure systems have a short shelf life. However, this is not a problem if the coagent is withheld from the stock during storage and then added in a short mixing cycle immediately before the stock is to be used.

7.3.2.5. Shared systems. Covulcanisation with highly unsaturated rubbers, which is an important capability that distinguishes halobutyl from other butyl rubbers, is achieved simply by using curing systems which are common to the halobutyl and the highly unsaturated rubbers present. For example, sulphur/accelerator systems, particularly those containing elemental sulphur, are used in blends of halobutyl and natural rubbers; zinc oxide systems are used for blends of halobutyl rubbers with other halogenated elastomers. Thiourea systems have been used for blends of neoprene with chlorobutyl, but they would be expected to scorch blends of neoprene with bromobutyl.

7.4. Processing
The mixing of halogenated butyl rubbers differs from that of regular butyls in that the halobutyls are too reactive to be heat t eated (they gel); zinc oxide and other curatives should not be added until late in the mixing cycle; stock temperatures should be kept below 135°C for bromobutyl, and 145°C for chlorobutyl, compounds; and residues of unsaturated

elastomers in the mixing equipment do not pose incompatibility problems. In other respects, halogenated butyl rubbers are processed by methods that are virtually identical to those described for regular butyl rubbers, except that stock temperatures during processing should be lower to allow for the very much greater reactivity of the halobutyls.

7.5. Vulcanisate Properties

7.5.1. General
Halobutyl and regular butyl rubbers provide equivalent gas impermeability, damping capability and resistance to flex cracking, the weather, ozone and various chemicals. However, halobutyl vulcanisates absorb water more readily, have lower compression set, and in some applications are much more heat resistant.

7.5.2. Heat Resistance
Halobutyl and butyl rubbers have virtually equivalent intrinsic heat resistance but cannot necessarily be compounded to perform equally well in any given high temperature application. In tyre curing bladders, for example, resin cured butyl rubbers retain their elastic properties much longer in steam at 150–200 °C than do the halobutyl rubbers. On the other hand, halobutyl tyre inner tubes for heavy transport vehicles survive much longer than butyl or natural rubber tubes at tyre running temperatures, and are much better in delaying the onset of surface degradation and consequent troublesome welding of the tube to the tyre carcass.

Curing halobutyl rubbers with zinc oxide alone, or with systems containing zinc oxide, produces excellent heat resistance. For example, in a study aimed at optimising bromobutyl heat resistance it was shown[19] that in terms of percentage of tensile product retained (tensile product = stress multiplied by strain at break), outstanding dry heat resistance is achieved with ZnO/ZDMC/MgO or ZnO/TMTD/MgO curing systems.

The same study revealed that antioxidants in general provide little if any additional stability to bromobutyl vulcanisates at temperatures below 150 °C, that many antioxidants contribute significantly to heat stability at higher temperatures, and that exceptional heat resistance is achieved with MBI/MgO, with diphenylamine/acetone reaction products plus MgO; or, particularly, with combinations of these two antioxidant systems. The author states that magnesium oxide was included with each system because, although its role in bromobutyl vulcanisation is not clear, there is no doubt that it contributes to processing safety, shelf life and heat resistance when used in conjunction with curing systems that do not contain elemental sulphur.

For optimum resistance to hot water and superheated steam, red lead (Pb_3O_4) or litharge (PbO) will perform better than zinc oxide (in curing systems in which the presence of lead is acceptable) because lead halides are insoluble in water and therefore inhibit water adsorption. Allowance may need to be made, especially in chlorobutyl compounds, for the lead oxides being less cure-active than zinc oxide.

7.5.3. Compression Set

A measure of the relative compression set capabilities of standard butyl, chlorobutyl and bromobutyl rubbers is given by the ranges shown in Table 15 for compounds having an optimum practical balance of compression set and cure time. The remarkably low set values achieved with the bromobutyl vulcanisate at this test temperature can be lowered even further by increasing the level of coagent used, if a substantially increased risk of scorch is acceptable.

TABLE 15

OPTIMUM COMPRESSION SET RANGES: ASTM B, 70 h AT 150 °C

Rubber cure system	Bromobutyl peroxide/coagent	Chlorobutyl TMTD/ZnO/MgO	Butyl resin
Set (%)	40–45	45–50	50–55

7.5.4. Water Absorption

Halobutyl vulcanisates usually contain small quantities of electrolytes. Consequently, they are less moisture resistant than butyl vulcanisates and are less suitable for use in electrical insulation. Nevertheless, as their structure restricts the thermal motion needed for diffusion, and as the amount of electrolyte present is small, they absorb moisture very much less readily than many other hydrocarbon elastomers. Consequently, they are capable of providing excellent service in such applications as hose tubes for hot water and superheated steam, and tank linings for aqueous solutions of corrosive chemicals.

7.6. Applications

7.6.1. General

Halobutyl rubbers are used in applications requiring butyl rubber behaviour but in which either standard butyl rubber cannot be used, or the halobutyl rubbers offer sufficient technical advantages over regular butyl to compensate for their higher cost. Thus, halobutyl rubbers are used in tubeless tyre inner liners, white sidewall and cover strip stocks, and blends

TABLE 16
INNER LINER—BIAS-PLY TRUCK TYRE

Formulation (parts by weight)

Polysar Chlorobutyl 1255[a]	75
SMR5—CV60[b]	25
Stearic acid	1
N-660 (GPF) carbon black	62·5
Naphthenic oil	12
Hydrocarbon resin tackifier	7
MBTS	0·75
Zinc oxide	5
Vultac[c]	1·2
Compound viscosity (ML-4' at 100 °C)	54
Δ Viscosity (168 h at 50 °C)	+3
Mooney scorch (t_5 at 125 °C) (min)	8
Mooney scorch (t_5 at 138 °C) (min)	5
Mill shrinkage (%)	34

Cured 8 min at 166 °C

Hardness, Shore A	51
Modulus at 100 % elongation (MPa)	1·5
Modulus at 300 % elongation (MPa)	6·5
Tensile strength (MPa)	11·5
Elongation at break (%)	510
Tear strength (kN/m)	
ASTM Die B	42
ASTM Die C	32
Brittle point (°C)	−44
Permeability to air ($Q \times 10^8$)	
At 35 °C	0·72
At 65 °C	3·2

Adhesion to 70 NR:30 BR light truck carcass (cured 30 min at 166 °C)

Unaged, measured at RT (kN/m)	3·4
Aged 48 h at 125 °C, measured at RT (kN/m)	2·4
Unaged, measured at 100 °C (kN/m)	3·0
Aged 48 h at 125 °C, measured at 100 °C (kN/m)	1·0

Aged in air, 70 h at 125 °C

Hardness, Shore A	57
Tensile strength (MPa)	9·8
Elongation at break (%)	310

Aged in air, 90 % RH, 720 h at 70 °C

Hardness, Shore A	50
Tensile strength (MPa)	6·5
Elongation at break (%)	480
Dynamic fatigue by extension cycling (megacycles)	
(extension ratio 2·0, 100 % elongation)	10

[a] CIIR, 1·2 wt % chlorine, 55 ML-4' at 125 °C, Polysar Ltd.
[b] NR, Standard Malaysian Rubber, 60 ML-4' at 100 °C.
[c] Alkyl phenol disulphide accelerator, Pennwalt.

TABLE 17
INNER LINER—RADIAL-PLY TRUCK TYRE

Formulation (parts by weight)	
Polysar Bromobutyl X2[a]	100
Stearic acid	1
MBTS	1·25
N-660 (GPF) carbon black	65
Pentalyn A[b]	7
Paraffinic oil	12
Zinc oxide	3
Sulphur	0·4
Compound viscosity (ML-4' at 100 °C)	55
Δ Viscosity (168 h at 50 °C)	+2
Mooney scorch (t_5 at 125 °C) (min)	21
Tel-Tak (kPa)	
To itself	250
To stainless steel	150
Cured 20 min at 166 °C	
Hardness, Shore A	55
Modulus at 100 % elongation (MPa)	1·0
Modulus at 300 % elongation (MPa)	3·3
Tensile strength (MPa)	8·2
Elongation at break (%)	780
Tear strength (kN/m)	
ASTM Die B	39
ASTM Die C	37
Aged in air, 240 h at 100 °C	
Hardness, Shore A	68
Tensile strength (MPa)	8·5
Elongation at break (%)	580
Aged in air, 336 h at 120 °C	
Hardness, Shore A	73
Tensile strength (MPa)	7·6
Elongation at break (%)	360
Adhesion to 100 % NR heavy truck carcass (cured 30 min at 166 °C)	
Unaged, tested at RT (kN/m)	25
Unaged, tested at 100 °C (kN/m)	17
Aged 336 h at 120 °C, tested at RT (kN/m)	6
Aged 336 h at 120 °C, tested at 100 °C (kN/m)	3

[a] BIIR, 1·9 wt % bromine, 52 ML-4' at 125 °C, Polysar Ltd.
[b] Tackifier, rosin acid ester with pentaerythritol, Hercules.

with highly unsaturated rubbers. They are progressively displacing butyl from pharmaceutical applications, tank linings, heavy-duty inner tubes, and various hose and belting applications. However, they are not being used in such butyl applications as light tyre inner tubes, chewing gum base, non-curing caulks and sealants, and tyre curing bags, in all of which butyl rubbers offer clear technical equivalence with, or advantages over, halobutyl rubbers, and are more economic.

7.6.2. Inner Liners
The functions of the inner liner in a tubeless tyre are to retain air in the tyre, to maintain air pressure at the correct level (which is important in terms of rolling resistance and tyre life) and to prevent air and moisture from penetrating the tyre carcass in amounts sufficient to cause damage to, or loss of cohesion of, the carcass components.

The properties required of a liner to perform this function include the ability to adhere permanently to the tyre carcass, heat resistance, flexibility and flex crack resistance over the full operating temperature range and a low level of permeability to air and moisture. Regular butyl rubber cannot be used in this application except, in the form of butyl reclaim, as a flexible, highly impermeable filler for other elastomers, because it cannot be made to adhere strongly enough to the tyre carcass. Bromobutyl rubber possesses all the properties, in ample measure, required for premium quality inner liners, but a few parts of natural rubber are normally included in chlorobutyl liner stocks to ensure that they have adequate cured bond strength (see Tables 16 and 17).

7.6.3. Heat Resistant Inner Tubes
Inner tubes made from standard butyl rubbers have a high level of heat resistance but they have been known to fail from heat softening (reversion) when exposed to the conditions under which inner tubes in heavy transport, bus and off-the-road tyres are required to perform. Halobutyl tubes under identical conditions are more likely to remain serviceable, as the heat ageing data in Table 18 show.

Bias-ply and radial-ply tyres require quite different inner tubes for optimum performance. Radial-ply tyres need:

(i) tubes with lower modulus $M_{300} \ngtr 4(MPa)$—to accommodate the greater strains on the splice,
(ii) greater splice strength—to withstand extensive flexing along the edges of the belt,

TABLE 18

HEAT RESISTANCE OF TYPICAL TUBE COMPOUNDS

	Bromobutyl	Butyl
Hardness, Shore A	45	53
Tensile strength (MPa)	7·8	11·4
Elongation at break (%)	570	670
Aged in air 22 h at 177°C		
Hardness, Shore A	63	60
Tensile strength (MPa)	4·1	0·3
Elongation at break (%)	590	40
Surface condition	dry	tacky
Aged in air 70 h at 177°C		
Hardness, Shore A	68	
Tensile strength (MPa)	1·9	(decomposed)
Elongation at break (%)	230	
Surface condition	dry	

(iii) greater abrasion resistance—because there is greater relative movement (squirming and scrubbing) between tube and tyre,

(iv) lower set—to prevent curing bladder venting channels and displaced cord ridges from imprinting, and thereby weakening, the tube wall.

Representative black and oil loadings reflecting these differences (for both butyl and halobutyl tubes) are shown in Table 19.

TABLE 19

TYPICAL BLACK AND OIL LOADINGS FOR BUTYL AND HALOBUTYL INNER TUBE COMPOUNDS

Bias-ply tyres	GPF (N-660)
Carbon black (phr)	60 FEF (N-550) ⟵⟶ 80 SRF (N-774)
	GPF . HS (N-650)
Oil (phr)	22 ⟵⟶ 30
Radial-ply tyres	
Carbon black (phr)	65 HAF . HS (N-347) ⟵⟶ 45–55 FEF (N-550)
Oil (phr)	30–35 ⟵⟶ 20–25

Typical formulations and representative properties for heat resistant inner tube compounds intended for use in bias-ply and radial-ply truck tyres are shown in Tables 20 and 21. Crosslinked butyl is included in the radial tyre tube to provide green strength. Butyl reclaim could be used instead, for the same purpose.

TABLE 20
INNER TUBE—BIAS-PLY TRUCK TYRE

Formulation (parts by weight)		
Polysar Bromobutyl X2[a]	100	
Stearic acid	1	
N-650 (GPF-HS)	70	
Naphthenic oil	25	
Zinc oxide	3	
MBTS	0·5	
ZD$_m$C	0·1	
Sulphur	0·5	
	Laboratory	*Factory*
Compound viscosity (ML-4' at 100 °C)	55	54
Mooney scorch (t_5 at 125 °C) (min)	15·5	18
Mooney scorch (t_5 at 138 °C) (min)	7·5	8
Extrusion		
Extruder	No. 1 Royle	23·6 cm Tuber
Screw speed (rpm)	30	30
Temperature (°C)	130	58
Die	Garvey	Tube (10·00–20)
Rate (m/h)	157	229
Die swell (%)	101	—
Appearance	A10	—
Cure time[b]		
(Minutes at 165 °C)	5	—
(Minutes at 170 °C)	—	3
Hardness, Shore A	43	45
Modulus at 100% elongation (MPa)	0·98	0·98
Modulus at 300% elongation (MPa)	6·1	5·0
Tensile strength (MPa)	10·7	9·0
Elongation at break (%)	520	530
Aged in air 10 days at 125 °C		
Hardness, Shore A	70	64
Modulus at 100% elongation (MPa)	4·4	4·0
Tensile strength (MPa)	10·9	8·8
Elongation at break (%)	220	290
Cured splice properties		
Strength (% of tensile strength)	—	90
Dynamic fatigue (kc to failure)		
(0–100% extension, 6 Hz, 70 °C)		
Unaged	—	57
Aged 48 h at 121 °C	—	25

[a] BIIR, 1·9 wt% bromine, 52 ML-4' at 125 °C, Polysar Ltd.
[b] Cure time for the laboratory stock is the optimum, as determined by Monsanto rheometer (t_{90}). Cure time for the inner tube is the shortest time required for freedom from porosity under the valve patch. Cure time for the same size (10·00–20) inner tube made from an identically loaded, TMTD/MBT/sulphur cured, regular butyl (Polysar Butyl 301) compound is 1·5 min longer (i.e. 4·5 min instead of 3·0 min).

TABLE 21

Formulation (parts by weight)	
Polysar Bromobutyl X2[a]	80
Polysar Butyl XL-50[b]	20
Magnesium oxide	0·5
Stearic acid	1
N-347 (HAF-HS) black	55
Naphthenic oil	35
Zinc oxide	3
TMTD	2·5
Compound viscosity (ML-4' at 100 °C)	44
Mooney scorch (t_5 at 125 °C) (min)	12
Mooney scorch (t_5 at 138 °C) (min)	7
Garvey Die extrusion, No. 1 Royle, 104 °C, 30 rpm	
Rate (cm/min)	185
Die swell (%)	140
Appearance	A10
Cured 8 min at 165 °C	
Hardness, Shore A	36
Modulus at 100 % elongation (MPa)	0·8
Modulus at 300 % elongation (MPa)	3·5
Tensile strength (MPa)	9·1
Elongation at break (%)	600
Tear strength (kN/m)	
Die B	40
Die C	26
Tension set, Smithers (%)	16
Tensile set (%)	20
Compression set, ASTM B (%) (70 h at 121 °C)	58
Goodrich flexometer	
Heat build-up (°C)	31
Aged in air, 168 h at 150 °C	
Hardness, Shore A	69
Modulus at 100 % elongation (MPa)	2·2
Tensile strength (MPa)	5·0
Elongation at break (%)	280
Aged in air 70 h at 175 °C	
Hardness, Shore A	63
Modulus at 100 % elongation (MPa)	1·3
Tensile strength (MPa)	1·5
Elongation at break (%)	200
Splice properties	
Tenacity (min)	>25
Cured strength (% of tensile strength)	75
Dynamic fatigue, cured tubes	
(kc to failure, 0–150 % extension 70 °C)	
Unaged	126
Aged 48 h at 121 °C	31

[a] BIIR, 1·9 wt % bromine, 52 ML-4' at 125 °C, Polysar Ltd.
[b] Crosslinked IIR, 50 % soluble in cyclohexane, Polysar Ltd.

TABLE 22
INNER TUBE—GIANT SIZE, OFF-THE-ROAD TYRE

Formulation (parts by weight)	
Polysar Bromobutyl X2[a]	50
Natural rubber	50
Stearic acid	2
Magnesium oxide	0·3
Antioxidant	1
N-774 (SRF-HM) black	60
Naphthenic oil	17·5
Treated whiting	12·5
Zinc oxide	4
MBTS	1·1
TMTD	1·65
Compound viscosity (ML-4′ at 100 °C)	35
Mooney scorch (t_s at 125 °C) (min)	13
Mooney scorch (t_s at 138 °C) (min)	7
Garvey Die extrusion, No. 1 Royle, 104 °C, 30 rpm	
Rate (cm/min)	137
Die swell (%)	154
Appearance	A10
Viscosity change	−2
Green strength, across grain (MPa)	
(100 % modulus)	1·1
Cured 10 min at 165 °C	
Hardness, Shore A	51
Modulus at 100 % elongation (MPa)	1·1
Modulus at 300 % elongation (MPa)	4·5
Tensile strength (MPa)	10·4
Elongation at break (%)	580
Tear strength (kN/m)	
Die B	53
Die C	37
Tension set, Smithers (%)	8
Tensile set (%)	12
Compression set, ASTM B (%)	
70 h at 100 °C	45
70 h at 125 °C	52
70 h at 150 °C	71
Goodrich flexometer	
Heat build-up (°C)	27
Aged in air, 240 h at 125 °C	
Hardness, Shore A	66
Tensile strength (MPa)	6·6
Elongation at break (%)	370
Aged in air, 22 h at 175 °C	
Hardness, Shore A	71
Tensile strength (MPa)	1·5
Elongation at break (%)	300

[a] BIIR, 1·9 wt % bromine, 52 ML-4′ at 125 °C, Polysar Ltd.

TABLE 23
WHITE SIDEWALL

Formulation (parts by weight)	
Polysar Chlorobutyl 1255[a]	30
EPDM	20
Pale crepe	50
Talc	20
Titanium dioxide (anatase)	50
Hard clay	30
Magnesium oxide	0·5
Stearic acid	1
Ultramarine blue	0·1
Zinc oxide	5
Vultac 5[b]	1·25
MBTS	0·75
Sulphur	0·5
Compound viscosity (ML-4' at 100 °C)	42
Mooney scorch (t_5 at 138 °C) (min)	8
Garvey Die extrusion, No. 1 Royle, 30 rpm, 104 °C	
Rate (cm/min)	235
Die swell (%)	79
Appearance	A6
Cured 30 min at 166 °C	
Hardness, Shore A	51
Modulus at 100 % elongation (MPa)	1·1
Modulus at 300 % elongation (MPa)	3·3
Tensile strength (MPa)	8·3
Elongation at break (%)	560
Ozone resistance (50 pphm, 40 °C, 168 h)	
Dynamic (0–25 % ext., 32 cpm)	no cracks
Static (20 % ext., 24 h relaxation)	no cracks
Flex crack growth, Demattia	
(kc to 300 % crack growth)	
Unaged	375
After ageing 2 weeks in air at 70 °C	> 300
Static peel adhesion to NR/SBR carcass (kN/m)	
At RT	11·6
AT 150 °C	2·3
Aged in air, 2 weeks at 70 °C	
Hardness, Shore A	54
Modulus at 100 % elongation (MPa)	1·6
Tensile strength (MPa)	7·4
Elongation at break (%)	540

[a] CIIR, 1·2 wt % chlorine, 50 ML-4' at 125 °C, Polysar Ltd.
[b] Accelerator, alkyl phenol disulphide, Pennwalt.

TABLE 24
TANK LINING

Formulation (parts by weight)	
Polysar Bromobutyl X2[a]	100
Stearic acid	1
Polyethylene AC-617[b]	5
Precipitated hydrated silica	10
Barytes	75
N-991 (MT) black	30
N-550 (FEF) black	20
Petrolatum	8
Factice	10
Zinc oxide	5
Sulphur	0·5
MBTS	1
TMTD	0·5
Compound viscosity (ML-4' at 100 °C)	54
Mooney scorch (t_5 at 125 °C) (min)	12
Monsanto Tel-Tak (MPa)	
Tack	0·22
Stickiness	0·12
True tack	0·10
Cured 7 min at 165 °C	
Hardness, Shore A	51
Modulus at 100 % elongation (MPa)	1·1
Modulus at 300 % elongation (MPa)	2·7
Tensile strength (MPa)	7·7
Elongation at break (%)	680
Adhesion to mild steel (shotblasted, degreased,	
cured in open steam 60 min at 0·28 MPa, jaw	
separation rate 5 cm/min)	
Bond strength (kN/m)	
90 ° peel	6·5
180 ° peel	10
Failure	cohesive

[a] BIIR, 1·9 wt % bromine, 52 ML-4' at 125 °C, Polysar Ltd.
[b] Low molecular weight polyethylene, Allied Chemical.

Giant inner tubes for off-the-road tyres usually require 25 phr to 50 phr of natural rubber in the compound to provide the high level of green strength needed during their manufacture. Filler, oil and curing system are adjusted, as indicated in Table 22, to minimise reductions in impermeability and heat resistance.

TABLE 25
PHARMACEUTICAL CLOSURE

Formulation (parts by weight)	
Polysar Bromobutyl X2[a]	100
Calcined hard clay	100
Red iron oxide	0·5
Polyethylene	2
Paraffin wax	2
Zinc oxide	3
ZD_mC	0·2
Compound viscosity (ML-4' at 100 °C)	78
Mooney scorch (t_5 at 125 °C) (min)	10
Monsanto rheometer at 180 °C, 3° arc, 10 cpm	
Max. torque (dN.m)	28
Min. torque (dN.m)	9
t_2 Scorch time (min)	2·5
Cured 4 min at 180°C	
Hardness, Shore A	50
Modulus at 100 % elongation (MPa)	0·7
Modulus at 300 % elongation (MPa)	2·0
Tensile strength (MPa)	7·4
Elongation at break (%)	900
Tear strength, Die C (kN/m)	23
Compression set, ASTM B (%)	
24 h at 70 °C	41
70 h at 100 °C	56
Permeability to air ($Q \times 10^8$)[b]	0·32
Turbidity (Nephelos units)	0·81
pH (change from blank)	−0·1
Reducing substances (ml 0·01 N I_2)	0

[a] BIIR, 1·9 wt % bromine, 72 ML-8' at 100 °C, Polysar Ltd.
[b] Q = volume (cm³ at NTP) passing per second through a specimen of 1 cm² area and 1 cm thick, when the difference in pressure across the specimen is 1 atm (0·1 MPa).

7.6.4. White Sidewall

The combination of elastomers used in the formulation shown in Table 23 provides an essential balance of adhesion, low heat build-up, and excellent resistance to ozone, weathering and flex cracking. Ultra-white grades of talc and clay were used. Bromobutyl would scorch if substituted directly for the chlorobutyl in this formulation.

7.6.5. Tank Linings

Chlorobutyl and bromobutyl have excellent resistance to polar chemicals

TABLE 26
ENGINE MOUNT, EFFECT OF NR:BIIR BLEND RATIO

Formulation (parts by weight)			
Natural rubber	100	50	—
Polysar Bromobutyl X2[a]	—	50	100
Stearic acid	1	1	1
Zinc oxide	3	3	3
Antioxidant	0·5	0·5	0·5
N-774 (SRF) carbon black	30	—	—
N-330 (HAF) carbon black	—	30	—
N-219 (ISAF-LS) carbon black	—	—	30
Aromatic oil	15	—	—
Naphthenic oil	—	15	—
Paraffinic oil	—	—	15
Sulphur	0·3	0·2	—
TMTD	0·5	0·5	0·5
CBS	3·0	2	1·25
Tensile strength (MPa)	21·4	16·0	13·3
After 72 h at 121 °C (% retained)	58	90	83
After 72 h at 150 °C (% retained)	0	16	52
Compression set, ASTM B (%)			
22 h at 100 °C	28	21	19
70 h at 100 °C	40	32	29
Demattia cut growth (mm)[b] at 40 kc	broke at 10 kc	2	7
Goodrich flexometer			
Heat build-up (°C)	20	50	60
Permanent set (%)	1	5	8
Rheovibron tester at 3·5 Hz			
Rebound (%)	75	50	26
Damping constant (tan δ)	0·5	0·17	0·27
Dynamic modulus (Pa × 10^7)			
At 3·5 Hz	0·35	0·35	0·35
At 110 Hz	0·37	0·60	0·83

[a] BIIR, 1·9 wt % bromine, 52 ML-4' at 125 °C, Polysar Ltd.
[b] Cut growth reached a minimum (2 mm) with the 50:50 NR:BIIR blend.

and perform well in tank linings. Bromobutyl cures faster and gives greater bond strength than chlorobutyl. In the compound described in Table 24, tack can be increased further, if a 10-point increase in viscosity is acceptable, by substituting polybutene for petrolatum.

7.6.6. Pharmaceutical Closures

A simple but effective compound is described in Table 25. Bromobutyl has a high level of purity and provides the required cure rate, resilience, needle penetration and sealing properties. Polyethylene contributes to processing, surface finish and texture. The selected grade of filler is particularly appropriate to pharmaceutical applications.

7.6.7. Elastomer Blends

Halobutyl rubbers can be blended successfully with virtually any elastomer that cures at approximately the same rate when sharing a common curing system. This capability is put to good use when there is a need to impart some measure of butyl-like properties to non-butyl compounds, to modify halobutyl compounds with other elastomers, or to reduce the cost of compounds based on more expensive elastomers.

Introducing halobutyl into natural rubber formulations for motor vehicle engine mounts, suspension bushings, steering couplings, and exhaust system mounts, all of which must withstand operating temperatures of 120 °C or more, is one practical example. The halobutyl improves heat resistance, shock absorbency, and resistance to cut growth, ozone and the weather, at some cost in terms of heat build-up, permanent set and resilience (Table 26).

TABLE 27

INFLUENCE OF BIIR IN A SIMPLE CR COMPOUND

Formulations (parts by weight)		
Neoprene GRT[a]	100	80
Polysar Bromobutyl X2[b]	—	20
Magnesium oxide	4	3·2
N-774 (SRF) carbon black	30	30
Zinc oxide	5	4
Stearic acid	0·5	0·5
Specific gravity	1·39	1·31
Tensile strength (MPa)	·22·4	20·4
Swell in ASTM oil No. 3 (%)		
70 h at 100 °C	+103	+126
Compression set, ASTM B (%)		
70 h at 121 °C	72	66
Tensile strength retained (%)		
Air aged 70 h at 121 °C	55	63
Cost by weight	100	96
Cost by volume	100	90

[a] CR, general purpose grade, Du Pont.
[b] BIIR, 1·9 wt % bromine, 52 ML-4' at 125 °C, Polysar Ltd.
Other elastomers with which halobutyl rubbers can be blended and covulcanised include NBR, EPDM, SBR, PPO, and IIR.

Another example is the use of halobutyl in neoprene compounds as an economy measure. Table 27 illustrates the influence of bromobutyl content on compound cost, tensile strength, oil resistance, compression set, and specific gravity. Air ageing resistance is unchanged.

ACKNOWLEDGEMENTS

The author wishes to express his thanks to numerous colleagues for their expert comments and constructive suggestions, and to Polysar Ltd for permission to publish.

REFERENCES

1. Sparks, W. J. and Thomas, R. M. US Patent 2 356 128, 1939.
2. I.G. Farbenindustrie, British Patent 421 118, 1934.
3. Buckler, E. J. and Adams, R. J. *Transactions IRI*, **29**, February 1953, 17.
4. Crawford, R. A. and Morrissey, R. T. US Patent 2 631 984, 1953. US Patent 2 681 899, 1954.
   Morrisey, R. T. *Ind. Eng. Chem.*, **47**, 1955, 1562; British Patent 768 628, 1957; US Patent 2 816 098, 1957.
   Morrisey, R. T. and Weiss, H. J. US Patent 2 833 734, 1958.
5. Baldwin, F. P. and Schatz, R. H. *Kirk-Othmer Encyclopedia Chem. Technol.*, **8**, 1979, 470.
6. IISRP, *The Synthetic Rubber Manual*, 8th edition, 1980.
7. Edwards, D. C. and Sato, K. *Rubber Chem. Technol.*, **53**, March 1980, 66.
8. British Patent 714 907 (US Rubber Co.), 1952.
9. Storey, E. B. and Edwards, D. C. *Rubber Chem. Technol.*, **30**, January 1957, 122.
10. Dolezal, P. T. and Johnson, P. S. *Rubber Chem. Technol.*, **53**, May 1980, 252.
11. Edwards, D. C. *Rubber Chem. Technol.*, **39**, June 1966, 581.
12. Edwards, D. C. *Rubber Age*, **78**, January 1956, 550.
13. Briggs, G. J., Edwards, D. C. and Storey, E. B. *Rubber Chem. Technol.*, **36**, September 1963, 621.
14. Buckler, E. J., Adams, R. J. and Wanless, G. G. Low temperature performance of butyl inner tubes, *Proceedings 2nd Rubber Tech. Conf.*, London, 1948, p. 34.
15. Kumbhani, K. J. *Butyl/Polyolefin Blends*, Polysar Technical Bulletin, March 1979.
16. *XL Butyl in Preformed Sealant Tape*, Polysar Technical Bulletin.
17. *Polysar Sealants Handbook*, July 1979.
18. Walker, J., Dunn, J. R. and Robinson, J. R. The development and utility of peroxide curing systems in brominated butyl rubbers, presented to the *Int. Rubber Conf., Tokyo*, October 1975.
19. Timar, J. and Edwards, W. S. *Rubber Chem. Technol.*, **52**, May 1979, 319.

Chapter 7

SILICONE RUBBERS

R. J. Cush and H. W. Winnan

Dow Corning Ltd, Barry, Glamorgan, UK

SUMMARY

Heat curing types of silicone and fluorosilicone rubber are discussed, including the very latest types of liquid silicone rubbers designed for moulding and coating (e.g. wire and fabric) processes. The trend of the industry towards supplying silicone rubber in base form is described together with a number of new additives which improve the processing and/or vulcanisate properties.

The developments in processing of silicone and fluorosilicone rubber indicate the enormous improvements made since the early materials were introduced and how even non-milling, ready-to-use, silicone rubber has been developed by controlling the filler/polymer interaction.

A very high proportion of the usage of silicone rubbers is associated with electrical applications, particularly wire and cable for the electrodomestic appliance and automotive industries. The inherent properties of silicone rubber make it a natural choice in areas such as safety (flame retardant) cables and food contact applications. Examples of the latter uses are given. In addition reference is made to automotive hose, electrically conductive applications, dynamic seals and medical implants.

The trend towards the use of silicone rubber in consumer products (cars, domestic appliances, etc.) seems certain to expand its use. This has been made possible by the unique properties of silicone rubber combined with a new commercial and technical flexibility for the fabricator of finished parts.

203

1. INTRODUCTION

Silicone rubbers have emerged during the last 30 years as highly versatile elastomeric materials available in a variety of forms and used in almost every major industry. Based upon high molecular weight polymers of dimethyl siloxane they exhibit the properties and performance characteristics traditionally expected of silicone materials. Industry generally regards silicone elastomers as two distinct product types:

1. Heat vulcanising rubbers—usually semi-solid, gum-like materials in the uncured form, requiring traditional rubber processing techniques for compounding and manufacture of finished parts.
2. Room temperature vulcanising (RTV)—usually flowable liquids supplied in a ready-to-use form for such applications as potting and encapsulation, mould-making, building sealants, etc. These are generally not used to produce moulded or extruded industrial parts.

A third generation of products is currently emerging in the silicone industry. These are heat-curable liquid materials specifically designed for the automated production of moulded parts and supported extrusions. This technology is described later in some detail and could represent a significant advance in processing during the next decade.

For the purpose of this chapter we will not discuss the applications and technology of the RTV liquid materials as this generally falls outside the scope of rubber industry practice. Developments in the physics and chemistry of silicone elastomers has been covered in great detail elsewhere and recently reviewed by Warrick et al.[1] In this chapter an up-to-date treatment of compounding, processing and application technology relating to the commercially available heat curable products will be presented.

1.1. Silicone Rubber Types and Basic Composition

Industrial grades of silicone rubber are based upon polymers of dimethyl siloxane, usually with small quantities of other organic groups substituted along the polymer chain. The incorporation of vinyl substituents can improve both cure performance, resilience and compression set. The inclusion of phenyl groups gives improved low temperature flexibility and some increased resistance to radiation. Addition of trifluoropropyl groups to the polymer chain results in a speciality type of silicone elastomer commonly known as fluorosilicone. This type exhibits outstanding resistance to many oils, fuels and solvents whilst retaining the basic characteristics of silicone polymers.

The chemical structure of a typical gum polymer can be represented as follows:

$$\begin{array}{cccc}
& CH_3 & CH_3 & CH_3 \\
& | & | & | \\
\sim\!\!Si\!\!-\!\!O\!\!-\!\!Si\!\!-\!\!-\!\!O\!\!-\!\!-\!\!Si\!\!\sim \\
& | & | & | \\
& CH_3 & CH\!\!=\!\!CH_2 & CH_3
\end{array}$$

Crosslinking usually occurs with peroxide radicals removing a hydrogen atom from a methyl group, which itself becomes an active radical attacking either adjacent methyl or vinyl substituents.

With many synthetic elastomeric polymers, the strength properties obtained from a non-reinforced crosslinked polymer are very low and generally unsuitable for industrial applications. Silicones are no exception and although carbon black can be used for reinforcement, fine particle size fume silica is the usual choice for property enhancement. The incorporation of these highly surface-active silicas into silicone gums is a difficult process due to the rapid interaction between polymer and filler resulting in a pseudo-vulcanised mass. For this reason a variety of siloxane based filler treatments are generally used to control viscosity and other properties.

The incorporation of reinforcing silica and treatment is nearly always carried out by the silicone manufacturer. Silicone elastomers are therefore usually supplied either ready-to-use or as a silica reinforced base requiring certain additives to crosslink and modify properties or performance.

ASTM D1418 has classified silicone rubbers in types to identify side group substituents and their resultant properties:

> MQ: Rubbers produced from simple polydimethyl siloxanes
> VMQ: General purpose products containing vinyl groups
> PMQ/PVMQ: Phenyl containing elastomers for extreme low temperature performance
> FVMQ: Fluorosilicone rubbers for fuel, oil and solvent resistance

1.2. Characteristics of Silicone Elastomers

Most silicone polymers, whether in elastomeric, fluid or resinous form, have performance characteristics that no other commercial material possesses in quite the same way. The more important of these can be listed as follows:

1. Suitable for service at temperatures from −100 to 300 °C
2. Excellent resistance to environmental degradation brought about by oxygen, ozone, sunlight
3. Excellent non-stick/surface release properties
4. Non-toxic
5. Can be optically transparent
6. Good electrical properties
7. Chemical inertness
8. Gas permeability

Without doubt, the single most important feature of silicone elastomers with regards to commercial applications is the high temperature stability of the product. If the useful life of an elastomeric product at high temperature is defined as that period in which the product retains the ability to reach 50 % elongation without failure, then both long range and accelerated ageing tests indicate a lifetime of years where other elastomers would fail in days or weeks. Table 1 shows typical results measured on a silicone rubber specially formulated for high temperature operation. However, in practice the service life of silicone parts is not easy to predict by accelerated or laboratory tests as often the 'in-service' conditions may fluctuate considerably.

The mechanism of thermal degradation is also directly related to the amount of oxygen present in the immediate environment as well as the presence of oils, chemicals or steam. When silicone elastomers are heated above 200 °C in close confinement, polymer reversion may result in softening and a loss of elastomeric properties. Usually this effect can be

TABLE 1
CONTINUOUS OPERATING TEMPERATURE VERSUS
USEFUL LIFE

Temperature (°C)	Service life (h)
150	15 000–30 000
200	7 500–10 000
260	2 000
316	100
370	1

minimised by the correct choice of compounding ingredients and cure system, and in practice rarely causes a problem.

A further characteristic of silicone elastomers is the relatively small change in mechanical and electrical properties when operating at an elevated temperature, as compared to ambient. Many organic elastomers having superior physical properties when tested at room temperature, lose this particular advantage in high temperature environments. At 150 °C the tensile strength of natural rubber, EPDM and even fluorocarbon elastomers, can fall below that obtainable from modern high performance silicone products tested at the same temperature.

For electrical insulation applications silicone elastomers retain their excellent dielectric properties over a wide range of operating temperatures. Even when exposed to flame conditions, the elastomer will burn or degrade to a non-conducting silica ash which will often maintain the electrical integrity of a circuit. The extreme resistance of silicones to ozone is of particular benefit in areas where corona discharge occurs.

The effect of solvent on silicone elastomers is one of swelling and softening. As with most other elastomer types, the change is a reversible physical phenomenon should the solvent or swelling medium evaporate from the rubber part. For good resistance to swelling by solvents and fuels, it is usually necessary to specify a fluorosilicone product. However, VMQ types in particular have sufficient resistance to swelling in many lubricating oils and other chemicals to be used in a variety of sealing applications, both static and dynamic.

The physiological inertness of silicone elastomers has led to many applications in the medical/pharmaceutical field. When properly post-cured the product is odourless, tasteless and generally accepted as completely non-toxic. The absence of adverse tissue reactions has led to silicone elastomers being specified for implant parts in the human body.

2. DEVELOPMENTS IN MATERIALS

Silicone rubber suppliers have continued to upgrade their products (particularly the strength properties) during the last 10 years so that the combination of these tougher grades of silicone rubber, with the many other material innovations made during this period, offers the industry a formidable armoury of products. The most significant developments are discussed below and cover such innovations as conductive silicone rubber,

flame retardancy, integral mould release additives, anti-'bloom' technology, improved long-term resistance to oxidised oil, non-peroxide cured
silicone rubber, base rubber compositions and high 'green' strength
products.

2.1. Improved Strength Characteristics

The very first types of improved tear strength (35–40 kN/m on ASTM Die B
test piece) silicone rubber were based on phenyl-containing siloxane
polymers but these had very poor resilience and hysteresis properties. They
were followed in the late 1960s by a new generation of high performance
materials which overcame these basic shortcomings, but which had
substantially the same level of tear strength as their phenyl-based
forerunners. However, while these new materials (with their much superior
resilience and relative lack of sensitivity to peroxide type and level)
expanded the usage of silicone rubber, it was not until the mid-1970s that
the first materials to have approximately 50 kN/m tear strength were
available. A typical stress/strain curve of such materials is shown in Fig. 1.
Note the higher modulus of these latest high tear types without any loss of
ultimate elongation, bringing them more into line with many speciality

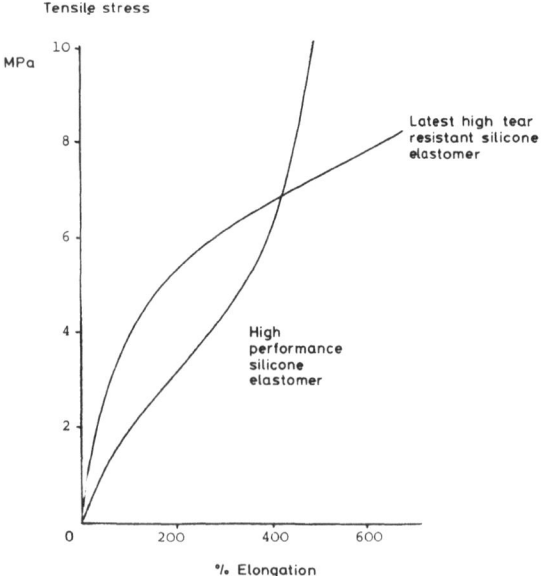

FIG. 1.   Stress/strain curve of silicone elastomers.

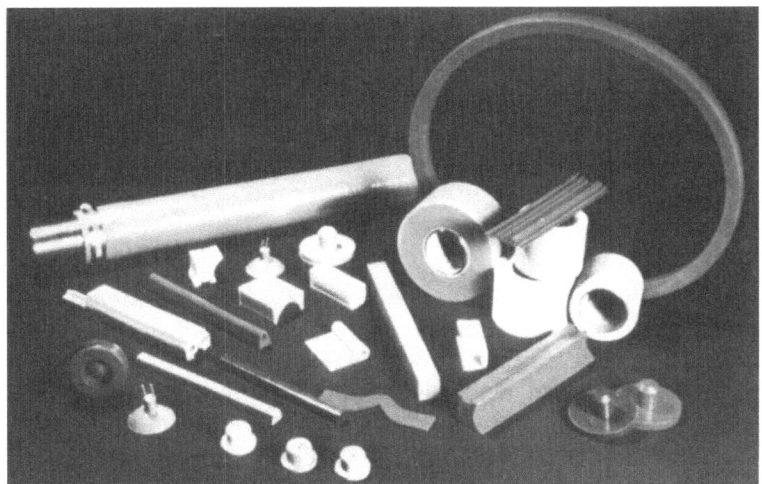

FIG. 2.   A selection of parts made from silicone.

organic rubbers, so that not only are measured values of tear similar to these organic rubbers but also they feel tough because of their high modulus. Such materials were initially hydrosilylation cured (SiH + SiCH=CH$_2$ addition reaction) which meant they were supplied in a two-pack form. Once mixed together in the prescribed ratio, usually 1:1, the materials had a shelf life of a few days before curing began, even at room temperature. Refrigeration of the mixture prolonged the shelf life to many months.

The use of hydrosilylation-curing technology also allowed silicone rubber to attain a better surface cure, particularly when hot-air vulcanised. The combination of a polished extruder die and such materials gave an outstandingly smooth, glossy finish to extrusions. Figure 2 shows a selection of parts made from this type of silicone rubber.

This very high level of tear strength (combined with tensile strengths of approximately 10 MPa) is now to be found in traditional peroxide cured products thus overcoming the limited shelf life problem of the mixed components referred to above.

2.2. Flame Retardancy
Although silicone rubber because of its basic inorganic nature burns less readily than most halogen-free organic rubbers, its burning characteristics are readily modified. The traditional halogen-containing flame retardant

TABLE 2
PRODUCTS OF COMBUSTION OF SILICONE RUBBER

Silicone rubber type	Gas concentration (ppm)					
	CO	HCl	HCN	HF	SO_2	NO
PVMQ	550	0	0	0	0	0
VMQ	350–800	0	0	0	0	0

additives can not only interfere with the peroxide (free radical) curing mechanism, but also greatly increase the toxicity of the smoke produced on burning. For this reason such additives are not normally used by silicone suppliers and have been superseded by equally or more effective additives, e.g. platinum compounds (sometimes in combination with titanium dioxide), carbon black, zinc or ceric compounds. Alumina trihydrate is also effective as a flame retardant in silicone rubber.

Typical general purpose grades of silicone rubber have a limiting oxygen index (ASTM D-2863) of approximately 25 (i.e. a mixture of 25 % oxygen + 75 % nitrogen is necessary to maintain combustion of a specified sample). By using the latest types of flame retardant additives offered by the silicone suppliers, values of over 35 have been recorded. In addition the low toxicity of the white smoke produced is maintained since there are no halogen- or nitrogen-containing constituents. The main products of combustion of silicone rubber are silicon dioxide (hence the white appearance of the smoke), water vapour and oxides of carbon. An analysis of the smoke produced when tested using the National Bureau of Standards smoke tests (flaming) is shown in Table 2.

2.3. Electrically Conductive Materials

Although the biggest single application for silicone rubber is as an insulator for electrical wiring due to its unique combination of high temperature resistance and electrical insulating properties, much work has been carried out to make electrically conductive or antistatic products. Such products are normally based on polysiloxane gum to which selected carbon blacks, such as acetylene black, are added. Reinforcing silica may be present also since carbon blacks have less reinforcing effect than silicas such as Aerosil. These compositions have been found to have volume resistivity values as low as 2 ohm cm. In addition they normally have a positive thermal coefficient of resistivity, i.e. their resistivity increases with increasing temperature, so that they are inherently safer in the event of electrical overloading.

Vulcanisation of conductive silicone rubber compositions is, like conventional silicone rubber, by peroxide. However inhibition of dichlorobenzoyl peroxide by carbon blacks makes it necessary to use an alternative curing technique, i.e. hydrosilylation or addition curing, for hot-air vulcanisation. Applications for conductive or antistatic silicone rubber include:

1. Stress cones in cable terminations
2. Calculator keyboard pads
3. Space heating panels
4. Antistatic rollers

2.4. Silicone Rubber in Base Form

Traditionally silicone rubber has been supplied to most processors in a fully compounded form, i.e. additives such as those for heat stability, improved processing, mould release, flame retardancy, etc. were already present in the formulation. In the late 1970s Dow Corning introduced their Silastic Compounding System as an alternative to the traditional form of supply. Essentially this system is a comprehensive range of base rubbers and modifiers which allow the user to formulate his own compositions in much the same way as the users of organic rubbers have done. This comprehensive range of products gives the user a degree of flexibility of choice not formerly possible. Some industries, such as the wire and cable industries, can be expected to continue to buy their silicone rubber in the traditional 'ready to use' form, as this is ideally suited to their continuous processes, e.g. cable extrusion. It seems likely, however, that the technical and cost flexibility described will help to enlarge this silicone rubber market at the expense of organic rubber, where proven cost/performance benefits are obtained.

2.5. High Green Strength

Silicone rubbers have Mooney viscosity values typically in the 35–50 range when measured at room temperature. This is why it is unnecessary to heat silicone rubber in order to obtain good flow characteristics during shaping processes. However, for processors more used to handling organic rubbers, silicone is softer and usually more prone to collapse—for example, in thin section extrusions.

In order to overcome this shortcoming high green strength silicone rubber is now available. It was first developed in the USA in order to allow hose to be made with silicone rubber by a continuous process originally designed for organic rubbers. Since then it has by virtue of its excellent handling characteristics found application in other extrusion applications. Improved green strength can also be imparted to conventional silicone rubbers by blending in a proportion of such a high strength grade, or more economically by use of a modifier from the Silastic Compounding System referred to in the previous section.

2.6. Integral Mould Release Agents

Silicone rubber is often moulded commercially without the need for any release agent, such is the low surface energy of the material. However, over a period of time mould fouling may occur, or, in the case of intricate parts made by injection moulding, it may be necessary to improve the demoulding characteristics. This has been done traditionally by spraying mould surfaces at regular intervals with release agents based on soap solutions, PTFE or certain silicone emulsions specially formulated for use with silicone rubbers. It is now possible to incorporate in the rubber a release agent which avoids the need for repeated spraying of the mould surfaces. At the levels normally used, approximately 0·25%, little or no effect has been found on other characteristics of silicone rubber.

2.7. Anti-'bloom' Additives

Hot-air vulcanisation of silicone rubber is usually only possible if 2,4-dichlorobenzoyl peroxide is used as the curing agent. The acidic decomposition products of this peroxide will 'bloom' to the surface of most silicone rubber compositions within one to two days after vulcanisation. For this reason it has been necessary to give a second (post) cure in an air-circulating oven at 200 °C or higher in order to remove these crystalline deposits. For many grades of silicone rubber physical properties, particularly compression set, were optimised by this post-cure.

It is now possible by incorporating acid acceptors, such as certain metal oxides, to prevent this bloom occurring. Normally less than 0·5% of a suitable metal oxide is found completely effective. However, such additives do not by themselves remove the need for a post-cure where compression set and heat stability properties are of paramount concern. In these cases a post-cure temperature should be selected at least 20 °C above the expected service temperature or compression set test temperature.

3. PROCESSING

Whilst conventional rubber industry techniques are normally used to compound and fabricate silicone elastomers, there remain some differences that result in silicone being an easier product to handle in many cases. As mentioned earlier, it is unusual for the compounder/fabricator to begin compounding with an unfilled silicone gum, as much of the proprietary technology of silicone manufacture is associated with the addition of silica, stabilisation of polymer/filler interactions and the incorporation at this stage of various reactive intermediates. It is more usual for silicone rubbers to be supplied already reinforced and stabilised in a base form requiring only additions of ingredients on a two-roll mill or, sold fully finished in a ready-for-use form. We will, therefore, concentrate on the processing of base and finished stocks as opposed to the techniques of incorporating reinforcing silicas.

3.1. Compounding

Although internal mixers can be used to add large amounts of extending fillers to silicone base stocks, the vast majority of compounding operations are performed on a two-roll mill. Many of the ingredients necessary to convert a base into a finished stock ready for fabrication, are available as pastes or masterbatches that only require mixing and cross-blending in a conventional milling operation. Mill mixing of silicones does, however, require some modification to the normal organic elastomer techniques.

As silicones are relatively soft materials at room temperature, often with little nerve or green strength, it is necessary to equip mills with close fitting cheek plates and a nylon faced scraper blade for removing the band of material during cross-blending and final sheeting (Fig. 3).

Unlike other elastomers, silicones are mixed and formed at room temperature. It is not necessary to heat the product to reduce viscosity and kill nerve. Indeed with fast peroxide curatives it is desirable to keep the base temperature below 40 °C in order to minimise peroxide loss due to volatilisation or, in extreme cases, to prevent scorching. Water cooling is normally specified for both rolls. Roll speed ratios of between 1·2:1 and 1·4:1 are recommended and it should be noted that most silicones will transfer onto the fast roll after banding.

Many silicone bases and finished stocks will require an initial softening before further compounding or fabrication can occur. This is due to a phenomenon known as crepe ageing which is a slow build-up of structure between polymer and filler. Occasionally this structure will build-up to the point, with bases stored for some months, where crumbling will occur on initial milling. As these crumbs are fed back into a fairly tight nip they usually break down rapidly to form a smooth band of soft material ready for compounding or use in a moulding or extrusion operation. More

FIG. 3.   Typical two-roll rubber mill with features recommended for milling
silicone rubber.

recently developed grades of silicone elastomers are often described as 'non-milling' type products. This means they have been formulated to keep structure build-up to a minimum and usually require no presoftening before fabrication, even when stored for some months. Certain additives used during compounding can also achieve this effect.

3.2. Curing and Post-curing

Silicone and fluorosilicone elastomers that require heat for crosslinking generally use some form of peroxide as the curing agent. Non-peroxide addition cure systems are extremely efficient at high temperatures, but are normally only used in heat curable liquid systems (Section 5).

The free-radical crosslinking process using peroxides can be completed in as little as 30 s at high temperatures during extrusion, although moulding applications generally have cure times in excess of this (2–5 min is typical for production parts). Mould cycle times are of course dependent on part size, mould complexity and heat transfer. An important characteristic of silicones is their flat cure profile. Once cure is complete, further heating will not improve or degrade properties. They are also relatively insensitive to cure temperature between the normal limits of 110–300 °C. In other words, there is no narrow band of optimum cure conditions beyond which the nature of the vulcanisate is changed.

Having stated this, there is however a procedure often used with silicones known as post-curing. This involves heat treatment of moulded or extruded parts in a forced draught air oven for usually between 4 and 24 h at 200–250 °C. This extended exposure to high temperature, circulated air results in the removal of peroxide decomposition products and other volatile materials. It can also provide an optimum stability of the elastomeric part where only minimal changes in physical properties can be tolerated in service conditions at high temperature. In many general purpose silicone bases and stocks, post-curing is also used as a means of obtaining the required property profile, particularly modulus and hardness. A number of new products have been introduced in the past decade which are described as 'no post-cure' materials. This indicates that post-curing is not necessary to reach optimum or expected properties, but may still prove desirable when the application demands all volatile residues be removed or to improve certain properties, such as compression set.

As described earlier, recently developed 'no post-cure' silicones may also contain additives to prevent peroxide bloom occurring without the need to post-cure.

3.3. Moulding
There are three principal methods of moulding silicone elastomers:

1.   Compression
2.   Transfer
3.   Injection

Compression moulding is still probably most widely used for low volume, large or difficult mouldings, but transfer and injection techniques are increasingly finding favour where production costs need to be reduced by faster output or automation. Conventional rubber moulding equipment is used in each case and moulds designed for organic elastomers often perform satisfactorily with silicones. It should, however, be realised that mould shrinkage with silicones may be higher than other elastomer types, and post-curing can also result in a further size reduction from volatile loss. Overall linear shrinkage values (with post-cure) are typically 2–5 % depending on product choice and mould temperature.

Moulds should include provision for the release of entrapped air and closed-end flow paths should be avoided unless a small bleed hole is incorporated. Most silicone elastomers will effectively push the air from a mould cavity during filling, but air pockets entrapped in the preform may give rise to soft uncured spots in the moulded part. This can normally be overcome by 'bumping' the press several times before the build up to full. pressure. Where manual or automatic bumping is not possible (e.g. transfer or injection presses) greater care must be taken in mould design and preform preparation.

As silicone elastomers flow easily under pressure, the forces involved in moulding can be lower than other elastomer types. Many prototype or short production-run moulds can be made of relatively soft materials, though for normal production situations, hardened steel is preferred. Highly polished cavity surfaces may well prove adequate, but chrome plating is often specified.

3.4. Extrusion

The extrusion of silicone elastomers is similar in technique to organic rubbers, but there are certain important differences:

1. Extrusion takes place at room temperature—both barrel and die head are often water cooled. Increasing the temperature of the process only increases the likelihood of scorch, it does not improve extrusion speed or surface finish which is generally excellent with silicones.

2. Silicones generally have less green strength (see Section 2.5) than organic rubbers and should be cured immediately after extrusion on a continuous basis.

Silicone elastomers prove to be such easy materials to extrude that no elaborate extruder or screw design is required. Both single and multi-flight screws are used with an L:D ratio of around 10:1. However, this is not critical, nor is the compression ratio which may be between 2:1 to 4:1 and is achieved by varying the pitch of the screw. The abrasive nature of many fillers used in silicone rubbers usually necessitates wear resistant nitrided steel, or similar materials, for barrel and screw assemblies, particularly at high output speeds. A roller feed mechanism for preformed strip is virtually essential for trouble free automated feed. With conventional feed throat designs, silicone elastomers will often form a bridge above the rotating screw.

The general necessity for silicones to be vulcanised continuously at the point of extrusion presents no problem. With the correct choice of peroxide, cure can be obtained in less than 1 min for thin section profiles by passing through an infra-red heated unit or hot-air vulcaniser (HAV). Pressure is not necessary to suppress porosity and many simple heating techniques work satisfactorily. Both hot liquid baths and fluidised bed systems have been used in the past, but the efficiency of a well designed high temperature, infra-red unit with conveyor belt, has established this procedure throughout the industry. Continuous steam vulcanisation (CV) is, however, used to advantage for some wire and cable manufacture. The coated wire passes directly from the extrusion die into a steam filled tube operating at pressures up to 18 bar. Cure can take place rapidly with this technique (10–15 s) and very high line speeds with long CV tubes are possible.

The type of preform used to feed many extrusion and indeed injection moulding machines, has undergone a number of developments in the past 10–15 years. In the late 1960s a move towards pelletised silicone rubbers was popular. However, problems of agglomeration and a build-up of talc (used as a dusting agent) in the extruder head was largely responsible for the gradual phasing out of this technique in favour of segmented strip. This preform is produced either by the silicone manufacturer or specialist compounder as a part of the straining operation prior to packing. A ram or large screw extruder is used to produce a four-segment strip which is coiled into an appropriate weight. The fabricator simply decides whether to use one, two, three or all four segments to match the extruder needs.

3.5. Fabric Coating/Calendering

Calendering is quite widely used as a technique for producing both supported and unsupported thin sheets of silicone rubber. The procedure and equipment are almost identical to that used for organic elastomers, except heated rolls are not required to ensure smooth flow and a good finish.

The poor green strength of some silicone elastomers causes problems with unsupported sheet. This can be overcome by calendering onto a flexible substrate or cloth, which can be easily removed after curing, and by use of newer high green strength elastomer types. Fabric coating with silicone rubber using a calendering technique is generally straightforward, with good penetration due to the easy flow characteristics of silicones. Both three-roll and four-roll calenders are used in a variety of configurations. Vulcanisation is usually carried out in a steam autoclave, the calendered sheet being wrapped onto a hollow metal mandrel. However, hot-air ovens or continuous vulcanisation directly from the calender are used where circumstances permit.

An alternative technique for producing silicone coated fabrics involves the use of solvent dispersions or pastes. Silicone elastomers are easily dispersed in a variety of hydrocarbon or chlorinated solvents, although the latter do tend to react with certain peroxides and reduce the shelf life.

Highly polar solvents such as acetone and the alcohols are poor dispersion media for VMQ type silicones, but ketones are the best choice for fluorosilicone, which is not readily dissolved by aromatic types. Peroxide loss always occurs in solvent coating techniques from reaction with the solvent, volatilisation during the drying stage and deactivation by atmospheric oxygen when cured in thin films. To counteract this effect it is normal practice to double or triple peroxide levels at least compared to those used for moulding or extrusion.

The technique of dispersion coating depends on the purpose of the final article. Where very thin coatings are required or only impregnation of a fabric is needed, low viscosity dip baths are used. For thicker coats or where only a one-sided coating is desired, a variety of spreading techniques are used. Regardless of technique, it is essential to remove all solvent from the coated layer before curing commences to prevent porosity in the final vulcanisate.

Usually a forced draught drying zone operating below 80 °C is the first stage of any curing tower. Once the solvent is removed, cure should take place as rapidly as possible, with hot air or infra-red units operating at a temperature to suit the fabric substrate.

With all fabric coating techniques, whether calendering, spreading or dipping it may prove advisable to use a priming system to ensure good adhesion to the substrate. Very often mechanical keying is adequate for the end application, but a variety of commercially available silicone primers give excellent bond strengths to glass cloth, cotton and synthetic fibres. The fabric is usually pretreated with primer and dried before the coating process. Adhesion is obtained during the curing of the rubber.

4.   FLUOROSILICONES

Fluorosilicone rubber is usually based on a gum with the following basic structure:

$$
\begin{array}{ccc}
CF_3 & CF_3 & CF_3 \\
| & | & | \\
CH_2 & CH_2 & CH_2 \\
| & | & | \\
CH_2 & CH_2 & CH_2 \\
| & | & | \\
-Si-O-Si-O-Si- \\
| & | & | \\
CH_3 & CH_3 & CH_3
\end{array}
$$

Vinyl groups are usually introduced at very low levels and substitute for a methyl group in the structure shown. This allows a greater variety of peroxide curing agents to be used to crosslink the rubber as well as obtaining better compression set properties as in the case of dimethyl rubber.

4.1. Improvements in Properties

The first fluorosilicone rubbers were introduced by Dow Corning in 1957 in order to fill the need for a solvent resistant silicone rubber principally in strategic applications where resistance to high and low temperatures is necessary. Fluorosilicone rubbers when first introduced were only available with a general purpose level of properties, i.e. tensile strength approximately 6 MPa, tear strength (ASTM Die B) approximately 15 kN/m. In addition the processing characteristics were poor, the rubber often sticking simultaneously to both rolls of a mill during the often lengthy process of breaking down any crepe formed during shelf ageing of the rubber.

Today's range of fluorosilicone rubbers was very recently increased by the addition of a second supplier, General Electric of the USA. Together with new products from Dow Corning which are vastly superior in processing to the early grades, fluorosilicones are now similar to the conventional VMQ types in processing and other characteristics:

1. High strength (tensile strength > 8 MPa; tear (Die B) > 30 kN/m)
2. High modulus
3. Base suitable for further additions of reinforcing silica, etc.
4. 40 Shore A and 70 Shore A blendable grades
5. Improved heat ageing
6. Improved reversion resistance
7. Increased resilience

4.2. Property Retention

Of course, fluorosilicone rubber is by no means the only solvent resistant rubber available, but it does have an extraordinarily good upper and lower temperature capability as well. Also a feature often overlooked by designers is the good retention of its properties when tested at elevated temperatures. This should not be confused with heat ageing performance in the conventional sense where material is subjected to a specified temperature for a particular time, after which it is cooled to room temperature and tested. This procedure tells us nothing of the properties of the material at the specified temperature and very little therefore about the mode of failure or performance of the material in service.

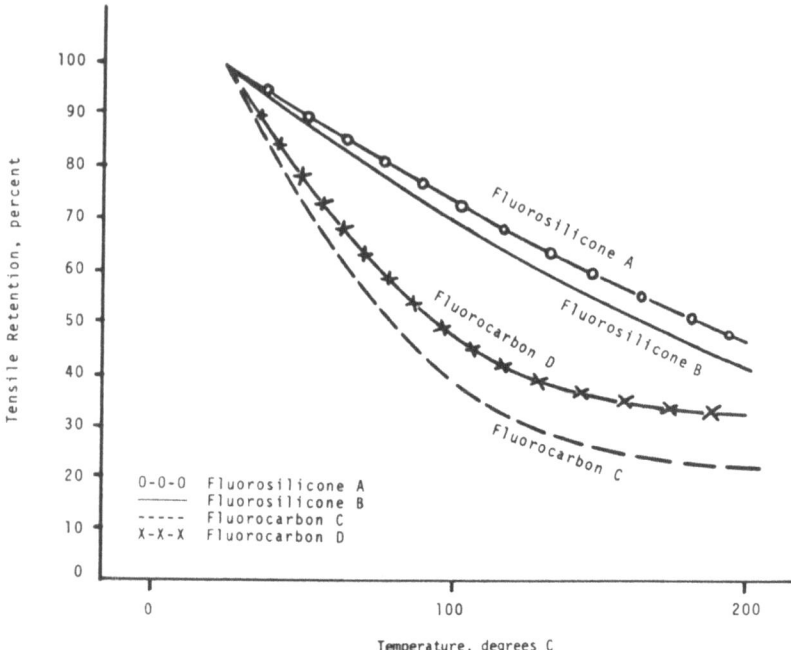

FIG. 4. Retention of tensile strength of selected elastomers at different
temperatures.

Let us take, for example, a comparison of fluorocarbon rubbers with
fluorosilicone on this basis. Commercially available fluorosilicone and
fluorocarbon rubbers were tested first at 25 °C and then at elevated
temperatures by the use of a temperature conditioning cabinet fitted to a
tensometer. Tensile strength and elongation at break were measured. The
results are illustrated graphically in Figs 4 and 5.

At 100–150 °C the two fluorosilicone rubbers tested retain 65 to 75 % of
their original tensile strength whereas the fluorocarbon rubbers tested only
retained 25 to 35 % of their original values. Elongation at break of the
fluorosilicone rubbers is little affected at temperatures up to 200 °C; at this
temperature approximately 70 % of the initial elongation of the
fluorocarbon rubbers has been lost. In practice this means that at service
temperatures above 100 °C fluorosilicone rubber has higher actual tensile
strength and elongation at break than most fluorocarbon rubbers.

The excellent fluid resistance of fluorosilicone rubber is a well established
fact. Recent data on the newer grades of fluorosilicone rubber also show

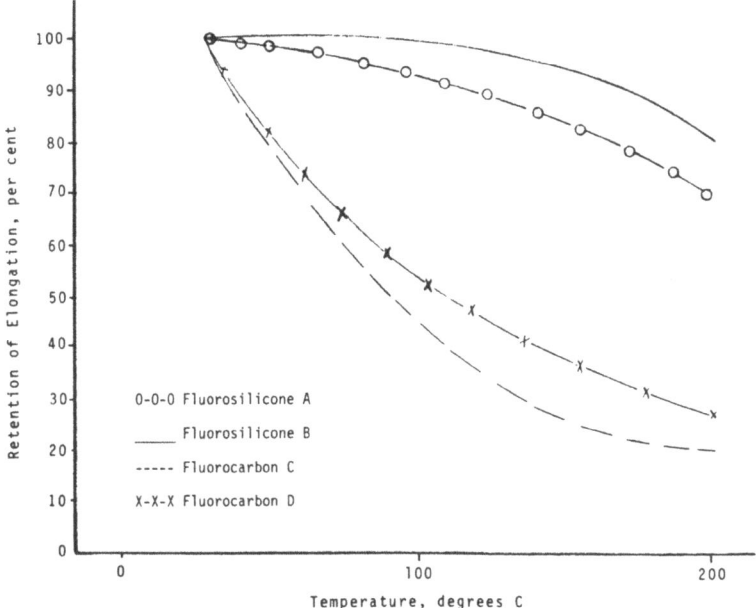

FIG. 5. Retention of elongation of selected elastomers at different temperatures.

that it is a very useful material where resistance to a mixture of gasoline and ethanol (viz. 'gasohol') is required. At a time when many countries in the world are evaluating or even using such mixtures as a way of conserving oil this means that the car designers are being forced to evaluate critically the materials used in fuel contact applications in their vehicles.

5. LIQUID SILICONE RUBBER (LSR)

The use of heat curable liquid silicone rubbers to manufacture a range of supported extrusions and moulded parts is a recent innovation in the development of silicone elastomer processing. For some years, addition cured RTV silicones have been available as heat curable liquid materials, but have never been widely used in this way for a number of reasons:

1. Short pot life
2. Limited physical properties
3. Limited heat stability

A new generation of LSRs has recently been commercialised which has been purpose designed for the manufacture of injection moulded and extruded parts by a fully automated process. These new materials overcome the earlier problems in using RTVs and result in finished parts with properties and temperature performance similar to conventional peroxide cured elastomers.

The benefit of this liquid processing system is mainly economic, resulting from lower production costs, increased output and reduced capital investment. There are, however, certain applications where a liquid system has practical as well as economic advantages mainly resulting from the low pressures necessary to form the product prior to curing.

5.1. Characteristics of LSR

Liquid silicone rubbers are low viscosity materials ranging from a flowable consistency to that of a thixotropic paste. All are easily mixed and delivered to the point of fabrication using liquid pumping techniques. The products are supplied fully compounded and de-aired, usually as a two-pack system which requires simple blending in a 1:1 ratio before use. This avoids much of the mixing and formulating costs associated with traditional elastomer processing. Pigments and any modifiers can be added continuously during the mixing stage, and this allows a wide range of colours from neutral or transparent base products.

Using an addition cure mechanism, LSRs exhibit rapid cure once the mass of rubber reaches 110 °C. In comparison with the fast peroxide cures (dichlorobenzoyl peroxide), completion of cure occurs in much the same time, but with a longer induction period and faster rate. This non-peroxide system allows LSR to be injection moulded at high temperatures (200–250 °C) with cure times of a few seconds for small parts. Scorch rarely presents a problem as the mould is quickly filled by the flowable liquid, even with very low injection pressures. Typical cure times for LSR as measured on a 12 g part in a Monsanto rheometer are shown in Table 3.

With LSR products post-curing is usually unnecessary as no by-products are formed during cure. Post-curing can, however, be used to advantage in certain applications where very low compression set is needed or to modify hardness/modulus. The use of addition cures is of particular benefit in hot-air vulcanisation techniques, where the use of peroxide always leaves a slightly undercured surface. LSRs give a non-tacky high gloss finish when cured in this way.

As a two-component system with 1:1 mixing of Part A and Part B, LSRs are designed so that a range of properties can be achieved from a limited

TABLE 3
CURE TIME OF A TYPICAL LIQUID SILICONE
RUBBER VERSUS TEMPERATURE

Temperature ($°C$)	Cure time
5	3 months
50	2 h
75	15 min
110	110 s
120	95 s
150	50 s
177	20 s
200	5 s

product line. The Part A of any product can normally be blended with the Part B of any other and some variation of the mix ratio is allowable for certain applications. The pot-life of mixed LSR is generally a few days at room temperature. If refrigerated to 5 °C or below this extends to weeks or months. However, even without cooling, there is usually no need to clean out equipment overnight, or even over a weekend. If longer shut-down is required then purging the system with one component only usually proves completely adequate for preventing cure at room temperature.

The physical properties of LSR are similar to the general purpose or medium high-strength peroxide cured elastomers. Other features such as oil and chemical resistance and electrical properties also follow the expected trend for silicone products. The LSR cure system generally results in the cured product being self-extinguishing, and with the inclusion of certain carbon black pigments will easily meet the requirements of the US Underwriters Laboratory flame test UL-94, with a class of V-O.

5.2. Mixing of LSR

A means of pumping, proportioning and blending LSR is essential for both moulding and extrusion applications. A range of commercially available meter-mix units will deliver mixed product under pressure to the fabrication process. Figure 6 shows a typical system in use for injection moulding.

Most meter-mix systems comprise three basic sections:

1. Removal of product from the shipping containers
2. Metering the 1:1 ratio
3. Mixing and delivery to the process

The units are pneumatically operated and consist of drum pumps, a

FIG. 6. Injection moulding with LSR.

proportioning unit and a static mixing device. A third component pump allows the addition of liquid pigments immediately prior to the static mixer.

The material components are mixed by passing them through a series of elements contained within the static mixing tube; these continuously divide the flow path to achieve a homogeneous mass without the inclusion of air, foreign matter or heat.

5.3. Injection Moulding of LSR

Injection moulding using LSR can offer significant improvements in process economics resulting from two basic considerations:

1. Lower labour/energy costs
2. Faster production of parts

Reduced capital investment and space requirements will also be a factor for new installations as compared with traditional mixing and moulding techniques. Injection moulding is chosen for LSR as this represents a rapid and economic way of producing a large number of small parts. This, being a fully automated process, produces items which can be manufactured, sorted and transported to packing containers without manual operation in many cases. For larger mouldings of fewer numbers, conventional semi-automatic transfer presses can be easily adapted to handle LSR.

If we consider a fully automated injection moulding process, specific benefits of using LSR can be:

1. No premixing or preforming is necessary. The meter-mix unit supplies direct to the injection moulding machine in a fully enclosed air-free system.
2. Low injection pressure is an important feature of an LSR system. Easy flowing liquids require only minimal pressure which can result in flash-free mouldings and use of a greater number of cavities per unit area of the mould. There is also less stress and wear of equipment and moulds.
3. Rapid cure at high temperature without scorch or distortion of the moulding. Cure times can be reduced to 25 % of that necessary for peroxide cured silicones formulated for injection moulding.
4. Little or no finishing or deflashing of parts.
5. No post-cure for many applications.

Most injection moulding machines can be modified to handle LSR. A few modifications are necessary, the number, depending on machine design. Whether a reciprocating screw or plunger-type machine is used, dynamic seals should be fitted to contain the liquid silicone rubber. This eliminates wasteful leakage of material and possible loss of injection pressure. A positive shut-off nozzle is necessary to isolate the injection unit from the mould cavity. This eliminates any drooling that may occur while the finished part is being removed from the open mould. On screw injection units it is important to have a check ring on the front of the screw to act as a one-way valve and ensure no back flow of LSR during injection. Because of the very low pressures used, clamp forces can be also kept to a minimum. Machines with clamps as small as 10 t are currently being used for production parts.

A number of injection moulding machines currently produced in Europe and the USA can be supplied already modified for LSR and provide, together with the meter-mix units, virtually a turn-key operation. Mould design is generally straightforward and duplicates many existing techniques used for rubber and plastics. Runner systems and gates can be much smaller than usual and this reduces material waste.

Liquid moulding techniques using LSR can reduce the process overheads of an operation to the point where even a relatively expensive raw material

FIG. 7.   Extrusion with LSR.

such as silicone can compete with speciality organic elastomers and give
better performance in the application.

5.4. Extrusion and Coating with LSR
At the present time the extrusion of LSR is confined to supported
applications, such as tapes and wire although unsupported extrusion
techniques are currently being developed.

Techniques of supported extrusion are basically similar to those of the
cross-head extrusion of rubber and plastic materials used by the wire and
cable industries. The main difference is the absence of an extruder! Liquid
silicone rubber is pumped directly to a cross-head from the meter-mix unit
or pressure pot. Curing is often carried out vertically using infra-red heaters
or circulated hot air, but horizontal HAV units can also be used. A typical
layout is shown in Fig. 7. The technology has been applied successfully to
wire coating, optical fibres, ignition core, various tapes and to braided glass

fibre sleeving. An adaptation of the extrusion technique can be used for cloth coating and spreading.

The advantages of a 100 % solids, heat-curable liquid system can be quite enormous compared to existing techniques where solvent dispersions are used. In all solvent containing processes it is essential to remove the solvent at a slow rate to avoid blisters in the cured article. This involves a low temperature drying stage which usually determines the overall speed of any continuous process. Using 100 % solids LSR the solvent removal restrictions do not exist and the coating line can operate as fast as curing allows. Thicker coats can also be applied in one pass. With the benefits of reduced fire hazard, no predispersing of elastomers in solvents and the in-line pigmentation, this technology could largely replace many existing techniques.

Wire coating with LSR can also offer a number of advantages compared to traditional methods. Less energy usage, improved surface cure and low extrusion head pressures are a few of the advantages being considered by the industry.

6. APPLICATIONS

The majority of applications for silicone rubber fall into one of the following categories:

1. Wire and cable
2. Pharmaceutical, food contact, para-medical
3. Seals and gaskets
4. Aeronautical and sponge
5. General extrusions and mouldings

6.1. Wire and Cable
Certainly the most important segment is wire and cable with over one third of the worldwide consumption of silicone rubber going into this application. The vast majority of such applications in Europe are associated with single core insulated wires used in electrodomestic appliances, such as electric cookers, irons, refrigerators, gas central heating ignition wiring, etc. A more critical application in this field is the use of silicone rubber in control and power cables for nuclear power stations.

In an extensive study[2] carried out by the Baltimore Gas and Electric Co., Baltimore USA, silicone rubber insulated cables of various designs were rated excellent and ultimately selected for the Calvert Cliffs Nuclear Power Plant in the USA. Clearly the safety aspects, flame retardancy, low toxicity white smoke, non-conductive ash formed on ignition, non-melting, etc. are paramount in such a potentially hazardous environment. Equally, silicone rubber is being used increasingly in other areas where safety is paramount. Examples include emergency lighting and alarm systems in high rise buildings, supermarkets, public buildings, North Sea oil platforms and underground railway fixed cables.

Silicone rubber insulated cables have been developed[3] which meet the stringent requirements of the IEC 331 specification, including the all important 3 h flame test. In this test the cable is required to function for 3 h at working voltage in a gas flame of 750 °C. In addition, silicone cable is available which meets BS 4066 and IEC 332 requirements for flame retardant cables. Silicone rubber also has a low calorific value (3·8 kcal/g) which when combined with the other features described above makes it a natural choice for cable insulation in many other applications.

6.2. Coolant Hose
Silicone rubber is well known for its chemical resistance and for its high and low temperature capability. These features have been recognised more recently by certain bus and truck operators who have replaced conventional organic rubber coolant hoses by silicone. A study[4] into the benefits of silicone hose on the Detroit Street Railway buses resulted in the changeover to silicone hose as far back as 1970. Although not widely used on cars as yet, silicone coolant hose, usually reinforced with polyester or polyamide fabrics, is well accepted for bigger vehicles, particularly those such as buses and trucks where 'off the road' time is expensive and frustrating. Thermostat by-pass hoses are available in the USA made from silicone rubber reinforced with nylon.

Turbocharger hoses in silicone rubber are relatively new and applications range from motor car engines to diesel locomotive engines.

6.3. Automotive
The most common uses of silicone rubber on European cars are:

1.   Crankcase seal (rear)
2.   Spark plug boots
3.   Ignition cables

The crankcase seal is a rotary shaft seal preventing the loss of oil from the crankcase via the drive shaft. The temperature at the lip of such seals is estimated to reach 180 °C due to frictional heat in service, and for this reason silicone rubber is one of the few materials with a good service life expectancy. This coupled with its excellent resistance to lubricating oils and flexibility at low temperature has made it the preferred polymer for many manufacturers.

Spark plug boots and ignition cables are used by the majority of the world's car producers, particularly in the USA where an all-silicone cable harness is commonplace.

Valve stem seals, cylinder liner gaskets, waterpump seals, bellows, exhaust emission control devices, carburettor needle tips, valves and diaphragms are other examples of where fluorosilicone or conventional silicone rubber are being used today.

6.4. Pharmaceutical/Food/Para-medical

National agencies concerned with regulating the materials for contact with food and drugs plus para-medical uses have long recognised the basic suitability of silicone rubber for such applications. The permitted silicone polymer types, fillers and additives are listed, for example in F.D.A. 21 CFR 177.2600 (USA) and Bundesgesundheitsblatt XV.III (Germany).

Silicone rubber, normally translucent, is used extensively for tubing on peristaltic pumps in hospitals and medical centres, and most commonly for pumping blood. The chemical and physiological inertness of such silicone rubber products, combined with their excellent pumping characteristics and sterilisability, are the reasons for using them in this application.

Similarly silicone rubber is used in many disposable hypodermic syringes as a plunger tip to ensure that no liquid escapes between the plunger and the walls of the syringe during use. In connection with the use of drugs small containers of injectable drugs often use a silicone rubber disc as a closure. Silicones may be made non-fragmenting so that after penetration of the closure by the hypodermic needle no silicone rubber fragments are deposited into the drug and on withdrawal of the needle the closure is self-sealing. Silicone rubbers which meet the stringent requirement of BS 3263 including non-fragmentability are commercially available. Closures for sterile water or saline solution bottles are similar examples.

In addition silicone rubber is being used for the following:

1. Tubing/seals in automatic vending machines for beverages
2. Baby bottle teats
3. Seals in food mixers and pressure cookers
4. Beer pump diaphragms

5.   Seals, etc. in coffee percolators
6.   Belts for deep freeze processing equipment and biscuit baking
7.   Confectionery moulds

Special grades of silicone rubber made under clean room conditions and rigorously tested for the content of specified elements, are also available commercially. These are used in the manufacture of implantable items in plastic surgery, and orthopaedic surgery, and for various parts such as catheters and drain tubes in urology, obstetrics, gynaecology and general surgery.

7.   FUTURE TRENDS

Silicone rubber in its various forms has been with us for nearly 40 years, during which time it has commanded a significant place in the spectrum of elastomeric materials. From the early days when the main applications were associated with military or strategic use we have seen the transition more recently into areas representing huge markets, such as cooker wiring, spark plug boots and rotary shaft seals. This trend into consumer product applications seems certain to accelerate and this is due in no small way to the unique combination of properties available in silicone rubber. Also a number of developments in materials and processing have added to the capability of such materials, whether it be from a processor's or user's point of view. In addition changes in the form in which silicone rubber is available, coupled with the option of buying modifiers (previously incorporated by the supplier) have increased enormously the technical and commercial flexibility available to the processor. It seems likely that the silicone rubber market will continue to expand into areas previously regarded as the prerogative of organic rubbers and that the year 2000 will find silicone rubber in such significant use that it will no longer be satisfactory for statisticians to include its consumption with 'miscellaneous synthetic rubbers'.

REFERENCES

1. WARRICK, E. L., PIERCE, O. R., POLMANTEER, K. E. and SAAM, J. C. *Rubber Chem. Technol.*, **52**, 1979, 437.

2. BHATIA, P. and BROWN, W. W. Flame propagation tests on 600 V control and power cables for Calvert Cliffs Nuclear Power Plant *I.E.E.E. Summer Meeting and International Symposium on high power testing*, 1979.
3. BENNETT, H. R. Cables for limited fire performance. *Electrical Times*, March 1977, 5.
4. ANON. Silicone hose, long-lasting economical in transit fleet. *Commercial Car Journal*, April 1971.

Chapter 8

SYNTHETIC POLYISOPRENE RUBBERS

M. J. Shuttleworth and A. A. Watson

Compagnie Française Goodyear, Centre Technique, Les Ulis, France

SUMMARY

Synthetic cis-*1,4-polyisoprene manufactured to a high degree of stereoregularity has become an important polymer for the rubber industry during the last decade. Assisted by considerable research and application development work, a selective usage pattern has emerged whereby the rubber is employed alone and in blends with other rubbers to exploit its outstanding processing qualities and its unique properties.*

It can be used in any application where natural rubber is the traditional choice and few, if any, compound changes are necessary. Changes in mixing and processing are required and result in significant time and power savings.

The major usage is in tyres, especially in critical compounds where its uniformity, quality and good flow properties allow the tyre constructors to meet the high standards demanded of steel-reinforced radial tyres. Engineering components such as springs and mountings form another major usage where its unique properties of low creep and compression set resistance are exploited.

Although modifications of the polymer have been attempted none, so far, have further improved performance of the polymer without detraction from its processing characteristics or vulcanisate properties.

1. INTRODUCTION

Synthetic *cis*-1,4-polyisoprene has been a commercial reality for 20 years and, despite complex economic factors related to the marketing of natural rubber and isoprene monomer feedstock price increases, production growth has been maintained and is likely to continue. 'High' *cis*-polyisoprene capacity worldwide now approaches 1 million tons per annum of which over half is in the USSR.

'High' *cis*-polyisoprene, either natural or synthetic, will remain an essential raw material for the rubber industry because of its combination of high strength and high resilience characteristics coupled with a broad-based utility. Therefore, it is important for rubber technologists to appreciate both the natural and synthetic polymers and to know how to use each to its best advantage.

Where 'high' *cis*-polyisoprene is required as the base polymer for the manufacture of rubber articles, either natural rubber or the synthetic counterpart can be used almost without exception.

The main objective of this chapter is to inform rubber technologists how and when to use synthetic 'high' *cis*-polyisoprene and to attempt, where possible, to link certain observed behaviour with studies made by rubber scientists concerning molecular structure. Emphasis is placed on situations where technical advantage is established for the synthetic polymer, since it is almost invariably sold at a premium to natural rubber and its value is appreciated by rubber manufacturers. Selective usage is the pattern that has developed and correctly so, since synthetic polyisoprene is unlikely to be a cheap replacement for natural rubber in the foreseeable future.

2. POLYMER STRUCTURE

2.1. Basic Types

One can discriminate between two basic types of synthetic polyisoprene. These depend upon the polymerisation catalyst system used and are commonly referred to as 'high' *cis*- and 'low' *cis*- types (Table 1).

The alkyl/lithium catalyst system, although capable of producing polyisoprenes having up to 98 % *cis*-1,4 structures under ideal conditions, has only been used commercially to give rubber with a *cis* content of about 92 %. This has limited its use and no other producer has followed the Shell Chemical Co. in commercially exploiting this process. For this reason greater attention is paid to the 'high' *cis*-polyisoprenes made using the Ziegler co-ordination catalyst systems of the trialkyl aluminium/titanium halide or poly(N-alkyliminoalane)/titanium tetrachloride types. The chemistry of polymerisation and the variety of structures which can be obtained have recently been extensively reviewed by our colleagues in the

TABLE 1
STRUCTURAL PARAMETERS OF POLYISOPRENES FROM INFRA-RED SPECTROSCOPIC AND
CHEMICAL ANALYSES

Polyisoprene (catalyst)	cis-1,4 (%)	trans-1,4 (%)	3,4 (%)	Head to head + tail to tail cis-1,4 (%)
Natural rubber	98	0	2	0
Natsyn 2200 SKI 3, Europrene IP 80 Nipol IR 2200 Ameripol SN 600 (Ti/Al)	96–97·5	0	2·5–4	2
Cariflex IR 309 (alkyl/lithium)	93	0	7	3

The appendix at the end of this chapter includes a list of registered trade names.

USA and reference to this publication[1] should be made by those requiring an in-depth knowledge of the subject.

2.2. Comparison with Natural Rubber

From the point of view of the physical characteristics of rubber compounds, their processing and vulcanisate properties, the significant differences between natural rubber and cis-polyisoprene and among different sources of the synthetic polymers relate to:

1. Stereoisomeric purity
2. Molecular weight and distribution
3. Presence of functional groups in the hydrocarbon chain
4. Non-rubber constituents

The structural parameters, determined by infra-red spectroscopic measurements (Table 1) show that the synthetic polyisoprenes produced using the Ziegler catalyst systems are closely similar to each other and are almost as structurally pure as natural rubber. Recent studies using the NMR technique[2] indicate that the natural rubber hydrocarbon is at least 99·6 % cis-1,4-polyisoprene. The small stereo-irregularities present in the synthetic polymers are sufficient to cause a reduced tendency for the synthetic polymers to crystallise either at low temperature or induced by applied strains. This difference in the rate of crystallisation, or perhaps the magnitude of the crystallites formed, is suggested to influence both processing and vulcanisate properties. The alkyl/lithium catalysed rubbers

TABLE 2
MOLECULAR WEIGHT DETERMINATIONS BY GPC

Polyisoprene	$\bar{M}_w$	$\bar{M}_n$	Distribution $\bar{M}_w/\bar{M}_n$
Natural rubber, RSSI	1 700 000	333 000	6·4
Natural rubber, SMR CV	1 140 000	185 000	6·2
Natsyn 2200	763 000	333 000	2·3
Nipol IR 2200	775 000	233 000	3·3
Ameripol SN 600	614 000	145 000	4·3
Europrene I P 80	569 000	165 000	3·5
SKI 3	770 000	230 000	3·3
SKI 3S	641 000	183 000	3·5
Carom 2230	330 000	98 000	3·3
Carom 2200	465 000	124 000	3·8
Cariflex IR 305	1 048 000	253 000	4·1

are significantly less stereoregular and, in consequence, show marked differences in physical properties compared to the 'high' cis-rubbers. Natural and synthetic polyisoprenes, as produced commercially, show differences in molecular weights and distribution (Table 2).

It may be misleading to relate these parameters too closely to processing and physical property differences for two reasons. Firstly, gel permeation chromatographic (GPC) examination of both natural and synthetic polyisoprenes can be carried out only on that portion (75–85 %) which is soluble. Secondly, during high shear mastication and mixing procedures, mechanico–chemical rupture of the hydrocarbon chains inevitably takes place which causes a large change in both average molecular weight and the distribution. Both natural rubber and 'low' cis-polyisoprene generally require premastication whereas the synthetic 'high' cis-polyisoprenes do not. A few aldehyde groups are present in natural rubber which can react with each other or with non-rubbers to form crosslinks.[3] This reaction can be blocked during production giving 'controlled viscosity' rubbers[4] which are marketed as SMR CV and LV grades by Malaysian producers.[5] Similar functional groups are absent in synthetic polyisoprenes.

Non-rubber constituents present in synthetic polyisoprenes are limited to small residues of catalyst and added stabiliser (BHT or similar phenolic antioxidant in non-staining types and DPPD + PBNA in SKI 3 and Carom 2230) and both ash and acetone extract contents are low, totalling 1–1·5 %. Natural rubbers generally contain non-rubbers, consisting of proteins, fatty acids and phospholipids, which total 3–6 %[6] depending upon the clonal origin and method of coagulation. These natural non-rubbers have small, positive effects on ageing properties and rate of cure but appear to be responsible for significant adverse effects on compression set and creep. They also cause reduced electrical resistivity compared to either deproteinised natural rubber (DPNR) or synthetic polyisoprenes.

Branching is believed to be present in both natural rubber and synthetic 'high' cis-polyisoprenes.[7] The higher degree of branching estimated for the natural polymer may be due to the few crosslinks present. The alkyl/lithium catalysed polymers are linear.[8]

3. RAW POLYMER PROPERTIES

3.1. Mooney Viscosity and Storage

Synthetic 'high' cis-polyisoprenes have Mooney viscosities of typically 75–80, although certain producers offer low Mooney grades having viscosities of about 60 which have proved useful in blends with other synthetic rubbers and to adhesive manufacturers.

In contrast to natural rubber, which suffers 'storage hardening' during transit from producing territories or on subsequent storage, 'high' cis-polyisoprenes retain their original viscosity. This gives the synthetic polymers the advantage of consistency over most natural rubber grades, except the viscosity-stabilised types which are becoming increasingly appreciated.[9]

Storage of natural rubber for long periods at ambient temperature, or for short periods at freezing temperatures, causes crystallisation of the rubber rendering it solid and unworkable. This necessitates use of 'hot-rooms' where the rubber has to be thawed for 24–48 h at temperatures of 40–50 °C. In our experience it is extremely rare for synthetic cis-polyisoprenes to crystallise in the bale form and, when it has occurred, thawing takes place in a few hours at 20 °C. Dilatometric measurements at -23 °C have shown that the rate of freezing is ten times slower for Natsyn 2200 compared to natural rubber (RSS 1).

3.2. Solution Characteristics

As a consequence of the lower molecular weight of 'high' cis-polyisoprene and probably reduced branching compared to natural rubber, dissolution in common solvents is both fast and provides solutions having significantly lower viscosities at the same concentrations (Fig. 1). This phenomenon has been put to useful purpose by some manufacturers of pressure sensitive adhesive tape by reducing the amount of solvent required in the adhesive

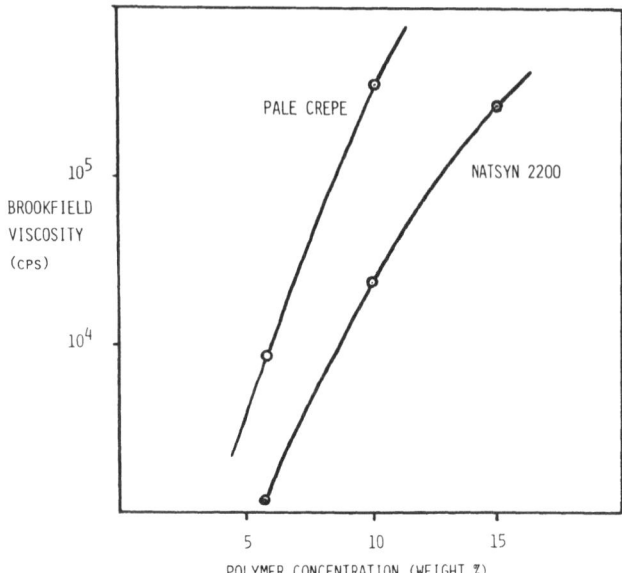

FIG. 1.   Solution viscosity of pale crepe and Natsyn 2200 at 23 °C (solvent: toluene, 60; isooctane, 40).

mass used for spreading onto the backing. Conversely, solvent adhesive manufacturers who sell their product by volume at a given viscosity find this characteristic distinctly unattractive.

4.   COMPOUNDING AND MIXING

In general, synthetic polyisoprenes may be compounded and processed using the same or very similar formulations and techniques to those used for natural rubber.

However, the rheological behaviour of the synthetic rubbers is such that significant changes in procedures can and should be made for optimum results. The most important factor is time and, almost invariably, substantial reductions in mixing cycles are achieved.

4.1.   Mixing Cycle Reductions

Synthetic 'high' *cis*-polyisoprenes largely lack the 'nerve' associated with natural rubbers so premastication, either as a separate mixing stage or at the start of a mix cycle, can be dispensed with.

TABLE 3

MILL[a] BANDING TIME AT 40 °C FOR NATURAL AND SYNTHETIC POLYISOPRENES

	SMR L	SMR CV	Natsyn 2200	Cariflex IR 305
Banding time, min	6	4·5	2·5	15

[a] Mill friction ratio 1 to 1·2.

This may be illustrated by comparing the times required to form a coherent band of rubber on an open mill (Table 3). The reason for this difference is due, at least in part, to the presence of a few crosslinks in general purpose raw natural rubbers formed during their preparation by the reaction of some aldehyde groups present in the polymer chain.[3] These aldehydes are not present in synthetic polyisoprenes.

In 'viscosity stabilised' grades of natural rubber, e.g. SMR CV, the aldehyde reaction is blocked by treatment with hydroxylamine during production and 'nerve' is thus reduced so that premastication can be minimised or eliminated during mixing operations.[10]

Nevertheless, some residual crosslinking reaction continues during the early stages of mastication. By following changes in Wallace rapid plasticity with mastication time, initial increases in viscosity are observed for natural polymers (Fig. 2). These differences account for the longer band formation times observed with natural rubber on the open mill.

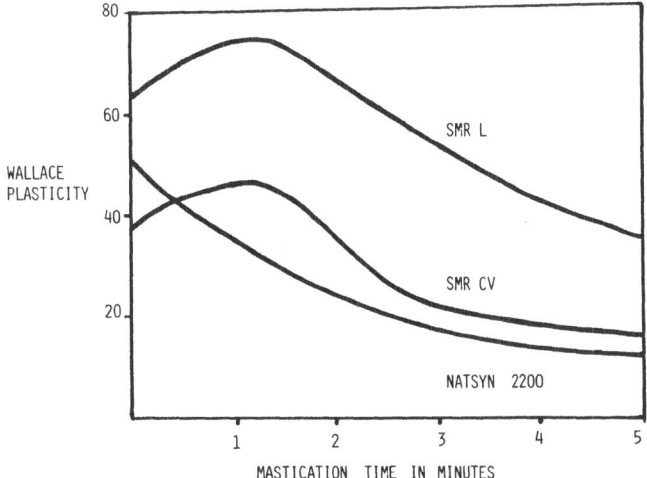

FIG. 2. Change in Wallace rapid plasticity with time during open mill polymer mastication (mill friction ratio 1:1·2, starting temperature 40 °C).

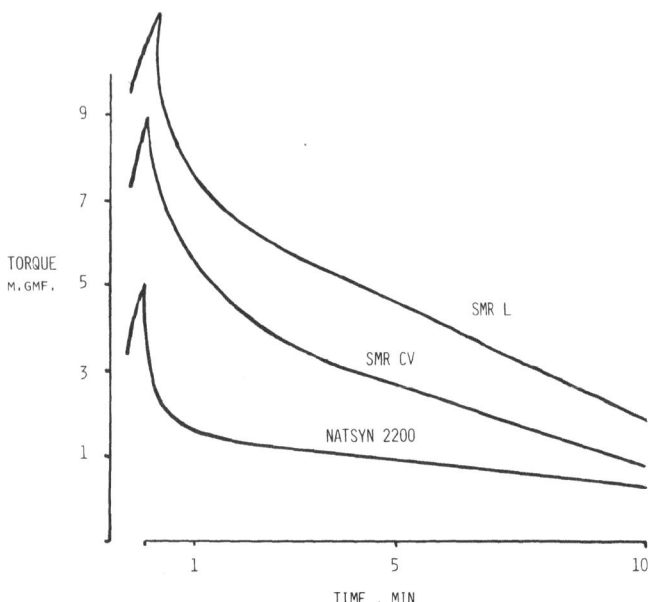

FIG. 3. Brabender Plasti-corder breakdown of polyisoprenes (speed 30 rpm, starting temperature 40 °C).

Different microstructure, molecular weight, molecular weight distribution and the presence of non-rubbers in natural rubber also account for changed behaviour compared to synthetic polyisoprenes and also between polyisoprenes themselves (see Section 2).

Raw polymer breakdown characteristics may be measured by observing Mooney viscosity changes from original values with time. This may be conveniently carried out using the Brabender Plasti-corder (Fig. 3). Although viscosity changes for synthetic polyisoprenes occur rapidly during the first few minutes of mastication, with continuing work-input relatively small changes occur. Natural rubber, by contrast, shows a greater and progressive viscosity reduction over a similar time period. As a consequence of this difference in rheological characteristics, acceptance of fillers by polyisoprenes must be rapid to ensure good dispersion while the polymer viscosity remains relatively high.

Using the Brabender Plasti-corder the black incorporation times for natural and synthetic polyisoprenes may be measured (Fig. 4). It can be seen that absorption and dispersion of filler is, as expected, much faster with the synthetic polymers.

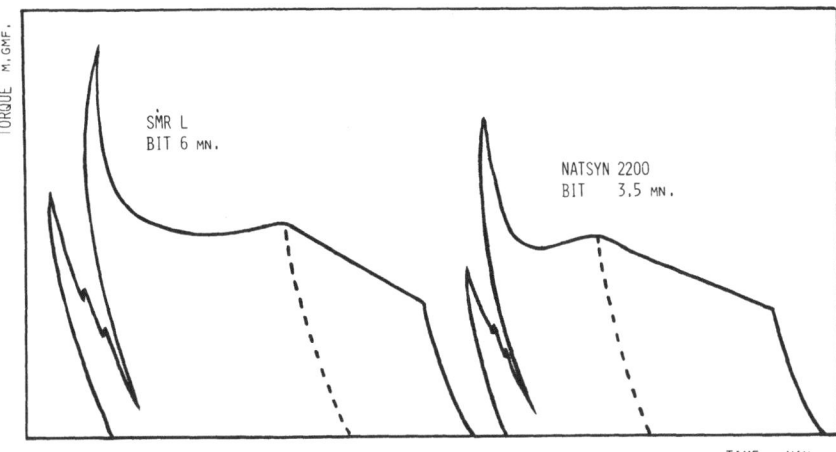

FIG. 4. Brabender Plasti-corder black incorporation times (BIT) for SMR L and Natsyn 2200 (polymer, 100; N-330 black, 50; speed, 100 rpm; starting temperature, 120 °C).

These observations demonstrate that, not only is it practically feasible to reduce premastication and mixing cycles to a minimum, but it is also technically desirable, to avoid unnecessary reduction in molecular weight of the rubber. If molecular weight reduction is excessive poor strength and resilience properties will result.[11]

4.2. Mill Mixing
Additions of compounding ingredients to synthetic polyisoprenes may be made as soon as a coherent band of polymer is formed on the mill. The sequence of ingredient addition is similar to that used for natural rubber. Chemically sensitive ingredients such as ultra-accelerators or blowing agents may be added without danger of prereaction, since even highly loaded compounds will mix at 10–15 °C cooler than comparative natural rubber compounds. Mixing cycle times may be reduced typically by 40 % with synthetic polyisoprenes whilst ensuring equivalent filler dispersion characteristics to natural rubber compounds.

4.3. Internal Mixing
The rheological characteristics of synthetic 'high' cis-polyisoprenes can be used equally to advantage in internal mixing operations. If polyisoprene is substituted for natural rubber at the same nominal batch weight, the batch

will have a lower effective volume in the mixer. The specific gravities of the rubbers are, of course, the same but voids are present in the chamber in the early stages of natural rubber mixing due to its 'nerve'. The ram will 'bottom' almost immediately after charging the mixer and less work will be done on the batch. To rectify this, synthetic polyisoprene batches should be increased by 5–10 % by weight by comparison with equivalent natural rubber batches.

Premastication of synthetic polyisoprenes is both unnecessary and in fact undesirable especially where incorporation of high structure carbon blacks is required, since high shear, obtainable only at the early part of the mixing cycle, is required to ensure good dispersion. Peptisers are rarely required.

For low and moderately loaded compounds, filler may be added directly after the polymer or, if equipment allows, over the ram. For highly loaded batches it may be necessary to split the filler addition to ensure rapid incorporation and minimum cycle times.

Internal mixing of synthetic polyisoprenes is also characterised by their ability to 'pick up' oils or other liquid ingredients or melted materials which tend to lubricate natural rubber batches and lead to wasted energy input and temperature drop within the batch.

As with open mill mixing, when using synthetic polyisoprene in internal mixers, cycle times are reduced and discharge temperatures may be up to 20 °C lower than those obtained with natural rubber, without sacrificing good dispersion of ingredients.

At the risk of labouring the point, it is the authors' experience that direct replacement of natural rubber by polyisoprene without increasing batch weight, without eliminating premastication or peptisers and by not reducing overall cycle times has led to very disappointing initial factory trials. Fillers have been badly dispersed and physical properties of vulcanisates have been poor due to excessive reduction of the molecular weight of the rubber. In an extended factory trial where the parameters mentioned above were closely controlled, 240 batches of compound having a 50 % replacement of natural rubber by polyisoprene were mixed in a No. 9 Banbury. Average mixing cycle times were 28 % less than for batches based on 100 % natural rubber and energy consumption was 15 % less. In addition these observations were made on batches which had a 3 % higher batch weight than the original all natural rubber compound. Typical power–time curves for these batches are shown in Fig. 5.

4.4. Compound Storage
Natural rubber–carbon black compounds have a tendency to increase in

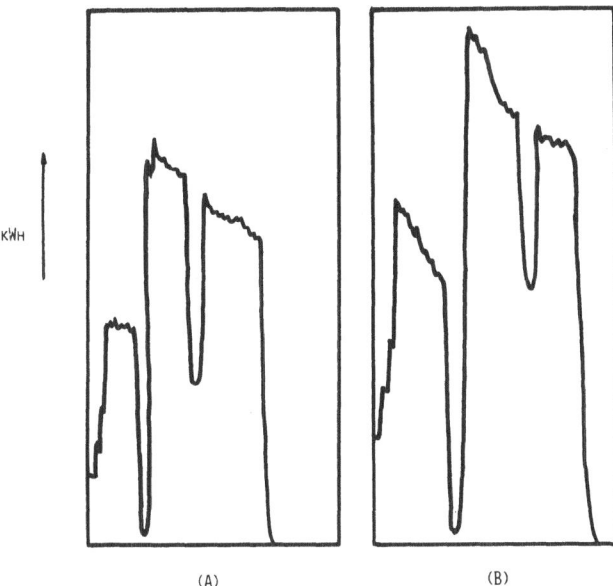

(A)   (B)

FIG. 5.   Power–time charts for masterbatch mixing (A) SMR 20, 50; Natsyn 2200, 50; N-330 black, 35; process oil, 3. (B) SMR 20, 100; N-330 black, 35; process oil, 3. (Banbury 40 rpm, starting temperature 55 °C).

viscosity on storage. The degree of viscosity increase will depend upon the grade of natural rubber, the type of carbon black and the duration and conditions of storage.

This phenomenon is recognised by the fact that such compounds need to be remilled or reworked prior to further utilisation, to ensure homogeneity and consistency of successive batches.

Synthetic polyisoprene batches exhibit far less viscosity increase on storage due to the slower crystallisation rate of the polymer. This may be illustrated by storing similar batches based on natural rubber and a synthetic polyisoprene and periodically measuring Mooney viscosities (Fig. 6).

4.5. Quality Control Parameters

Traditionally, natural rubber and natural rubber compounds are conveniently characterised for their consistency by Mooney viscosity measurements. The test is rapid and lends itself to factory usage. However, batches having similar Mooney viscosities but based upon different base

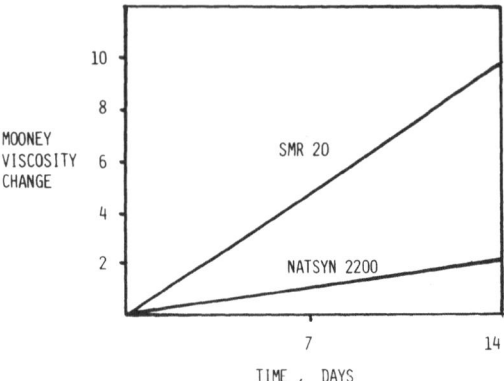

FIG. 6.   Mooney viscosity increase during storage of compounded rubber stocks (polymer, 100; N-660 black, 45; storage temperature, 20 °C).

polymers may not be expected to behave in the same manner during further processing operations.

The low shear Mooney apparatus does not predict the difference in flow behaviour between natural and synthetic polyisoprenes. Processing and shaping are relatively high shear operations and the differences between natural rubber and synthetic polyisoprene are much more pronounced during extrusion, calendering and moulding. It is normal for the Mooney viscosity of a black filled polyisoprene to be significantly higher than that of a natural rubber compound and yet be much easier to extrude, inject, calender or mould.

High shear testing machines such as the Wallace rapid plastimeter, the Monsanto processability tester and extrusion viscometers may better differentiate between natural rubber and polyisoprene, and even between polyisoprenes from different suppliers. None of these machines alone, however, will accurately predict the behaviour of the polymer or batch for all processing operations. Thus Mooney viscosity plus practical experience remain the principal factors in categorising factory processing.

5.   SECONDARY PROCESSING

5.1. Remilling
Rubber compounds are often remilled or remixed in the internal mixer after storage to ensure uniformity of product and reduce Mooney viscosity to a common and usually low level to facilitate subsequent shaping.

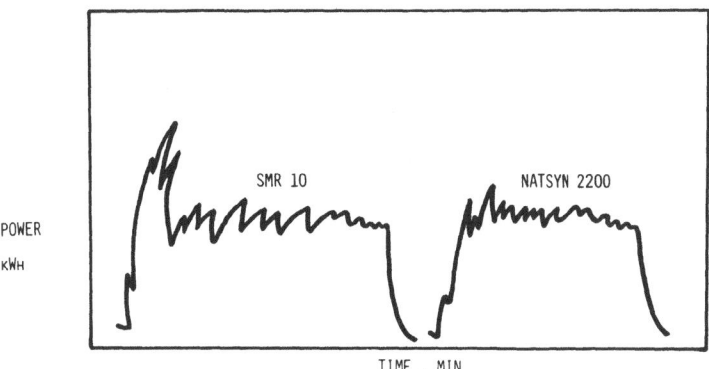

FIG. 7. Power–time curves from No. 9 Banbury for reworked rubber-black masterbatches showing shock loads.

As mentioned in the previous section, natural rubber compounds tend to harden and increase in viscosity on storage and reintroduction to an open mill or internal mixer creates shock loads and high power demand. This may be illustrated by examining the power–time curves of compounds during remilling (Fig. 7). After initial high power load, natural rubber requires considerable further power input to reheat and rework the batch. With synthetic polyisoprenes shock loads are less, due to the much reduced change in hardness and viscosity of black masterbatches during a storage period. Again due to flow properties, batches start remixing very rapidly and, while there is a reduction in viscosity at this stage, the change in viscosity is the same as in the finalisation (cure addition) stage. It has been established under factory conditions, that remills of polyisoprene compounds are of extremely limited value and that their elimination leads to gain in productivity and reduction of power costs.

5.2. Extrusion and calendering
For these processing operations rubber compounds undergo high shear deformation and the flow characteristics of synthetic polyisoprenes can be used to advantage.

Following the mixing operation polyisoprenes should have higher Mooney viscosities than equivalent natural rubber compounds but, nevertheless, will still exhibit faster flow under high shear. Compounds based on the synthetic polymer may be cold fed to extruders in which the necessary prewarming prior to extrusion is easily achieved or, alternatively, for hot feed extruders premilling should be kept to a minimum to avoid too

TABLE 4

DIE SWELL CHARACTERISTICS

Base polymer	Percentage swell at 30 rpm	Percentage swell at 75 rpm
SMR 10	17·5	18·7
Natsyn 2200	0·5	1·0
Europrene IP 80	1·0	3·4
Carom 2230	4·2	11·1
SKI 3 S	4·4	9·2

Formulation to ASTM D 3184–75–2A.
Extrusion on Brabender Plasti-corder.
Temperatures 70:70:100°C.

soft a stock entering the feed zone. Under identical extruder conditions, optimised for natural rubber, synthetic polyisoprene will extrude faster and with less die swell than the natural product (Table 4). In addition, at temperatures too low to achieve good extrusions with natural rubber, polyisoprenes will still give excellent definition and extrusion speed (Fig. 8). In either case, it is desirable to effect extrusion operations at the lowest temperatures consistent with the quality of the extrudate, since higher back

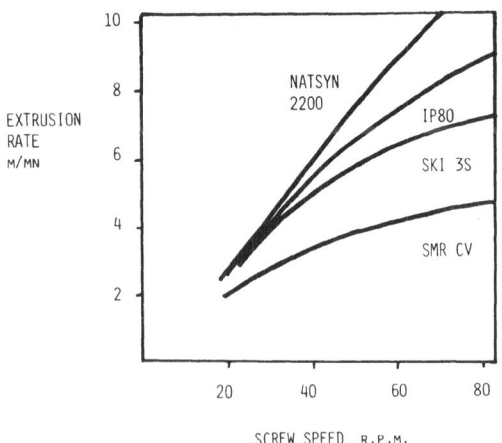

FIG. 8. Extrusion rate versus extruder screw speed for various polyisoprenes
(extruder temperatures 70:70:100°C).

pressure in the barrel will lead to higher output. Reduced warming on mills prior to extrusion, coupled with less heat generation in the extruder, also substantially reduces the tendency of sulphur to bloom from extrudates.

In calendering operations the same handling of synthetic polyisoprenes should be observed. The calendering of polyisoprenes is characterised by the ease of transformation into sheet or impregnation during frictioning operations together with a stability and reduced shrinkage in the former operation which cannot be achieved with natural rubber based compounds. Provided calender temperatures are kept 10–15 °C lower than for natural rubber there have been no situations encountered by the authors where 'green strength' deficiencies have been observed. In fact there are several known examples where very hard highly loaded natural rubber stocks with a tendency to scorch at the calender have, by inclusion of polyisoprene, been able to be run cooler and, thus, this problem avoided.

5.3. Injection Moulding

This processing operation gives rise to high shear effects in rubber stocks of at least an order of magnitude greater than in extrusion and calendering. Natural rubber compounds are thus usually specially compounded[12] to ensure the necessary scorch safety at the injection port during the actual injection cycle. Failure to do so may result in cured pieces of compound being injected into the mould.

Because of the high frictional heat build-up in natural rubber compounds at the injection phase, fast cycling by injection can be obtained. Synthetic polyisoprenes injected under the same machine conditions and with the same formulation as for natural rubber may be undercured, particularly in thick mouldings due to the lower frictional heat build-up in the compound during the injection phase. This effect can be used to advantage and overall cycle times reduced but requires some effort for correct optimisation of injection moulding conditions. Firstly, polyisoprenes should be run in the preheating zone of the machine at a temperature 15–20 °C higher to compensate for lower frictional heat build-up. The injection port temperature should also be higher to ensure that the compound is delivered into the mould close to the mould temperature. Finally the mould temperature may be increased to obtain faster cure. The generation of frictional heat at the injection port has a much greater effect on temperature build-up in the compound than external heating of the injection machine. Thus the preheating zone temperature increases described above still permit the use of very active cure systems in synthetic polyisoprene compounds which in natural rubber compounds would cause scorch or actual cure in the injection port. With lightly loaded or non-black compounds it may be desirable to use non-retarded cure systems with synthetic polyisoprenes for the same reason.

The generally accepted upper limit of temperature for moulding natural rubber compounds is around 180 °C above which there is a tendency to foul moulds. Due to their lower content of non-rubbers synthetic polyisoprenes may tolerate mould temperatures of 200 °C. The advantage of lower fouling tendency is evident in reduced down time of machines while moulds are changed for cleaning.

6. VULCANISATION

All the well-established systems of vulcanisation used for natural rubber can be used for synthetic polyisoprenes and usually little or no change is required in the vulcanisation recipe. Nevertheless, there are certain factors which are important and advantages to be gained by correct formulation.

6.1. Activators

Natural rubber invariably contains fatty acids which have an activating effect in accelerated sulphur vulcanisation systems. This type of activator is absent in synthetic polyisoprenes and must be added in sufficient quantity (1·5–2·0 phr) to ensure full crosslink development. Alternative activators of a similar chemical nature, such as zinc stearate or zinc 2-ethylhexanoate, have been used at lower levels (see Section 7).

Although proteins and their degradation products activate natural rubber cures, it is not necessary to incorporate materials such as soya bean lecithin into 'high' *cis*-polyisoprenes as has been recommended for the 'low' *cis*-polymers.[13]

In gum or mineral filled compounds, synthetic polyisoprenes exhibit longer times to optimum cure than natural rubber and significant differences between the various synthetic polymers are observed with some vulcanisation systems. The lack of naturally occurring amine activators present in natural rubber becomes evident in a simple thiazole-accelerated sulphur cure system. This is very well illustrated[29] by observing the incremental addition of diphenylguanidine (DPG) to a MBT-accelerated sulphur system (Table 5) where the lack of added amine accelerator shows a very slow cure for the synthetic polymer but, if sufficient is added, the differences between the rubbers become indiscernible.

TABLE 5
EFFECT OF ADDITION OF DPG TO AN MBT-ACCELERATED SULPHUR SYSTEM IN GUM NATURAL RUBBER (NR) AND A SYNTHETIC POLYISOPRENE (NATSYN 2200)

	NR	Natsyn 2200	NR	Natsyn 2200	NR	Natsyn 2200
DPG level, phr	0	0	0·25	0.25	1.0	1.0
Monsanto rheometer cure characteristics						
ts_2, min	22	34	13	15	5	6
$t'c$ (90), min	43	70	23	27	12	11·5
Max. torque, in lb	38	36	46	49	51	51

Base compound: Polymer, 100; stearic acid, 2; zinc oxide, 3; sulphur, 2; antioxidant, 1; MBT, 1.

6.2. Black-loaded Compounds

For black-loaded compounds, using conventional sulphur/mercaptobenzothiazole sulphenamide cure systems, 'high' *cis*-polyisoprenes show the following characteristics in comparison with natural rubber:

1. Greater scorch safety
2. Equal or slightly longer optimum cure times
3. Lower rheometer modulus development

Table 6 illustrates the cure behaviour of different polyisoprenes relative to natural rubber by comparison of Monsanto rheometer data.

It should be noted that all the synthetic polymers give longer scorch times. However, by comparing the differences between partial and

TABLE 6
SYNTHETIC *cis*-POLYISOPRENE CURE CHARACTERISTICS RELATIVE TO NATURAL RUBBER IN A CONVENTIONAL BLACK COMPOUND

	SMR 10	Natsyn 2200	Europrene IP80	Carom 2230	SKI 3S	SKI 3
Monsanto rheometer (150°C, 100 cpm 3° arc)						
ts_2, min	5	6·5	9	7	8	6
$t'c$ (50), min	8	9·5	13	11·5	12	11
$t'c$ (90), min	14	14	17	16	17	16
Max.–min. torque, in lb	64·5	58	47·5	47	48	53·5

Polymer, 100; zinc oxide, 5; stearic acid, 2; HAF black, 35; Santocure NS, 0·7; sulphur, 2·25.

optimum cure, one sees that some polyisoprenes exhibit a faster rate of cure and can give optimum cure times equal to natural rubber.

6.3. Crosslinking Efficiency and its Improvement
The differences between torque values developed by the various synthetic polyisoprenes parallel differences in modulus values and reflect cross-linking efficiency. The higher torque development for natural rubber compared to Natsyn 2200 is believed to be due partly to a different viscoelastic behaviour since, for example, hardness is higher for Natsyn 2200 than for natural rubber and estimates of crosslink densities by swelling measurements show a close similarity.

From the practical standpoint, two minor alterations in cure system should be considered if the cure characteristics of a synthetic polyisoprene need to be altered to match exactly that of a natural rubber compound. A secondary accelerator (e.g. 0·1–0·2 phr of TMTD) may be added to a sulphur/sulphenamide system to reduce time to optimum cure. An increase of between 5 and 15 %, *pro rata*, in both sulphur and accelerator will improve the degree of cure, as indicated by maximum torque value on a Monsanto rheometer.

7.   PHYSICAL PROPERTIES

7.1. General
'Low' *cis*-polyisoprene is generally used as a part replacement for natural rubber (limited to levels between 15 and 20 parts), the main objective being to facilitate processing. The comparative physical properties shown in Table 7 illustrate that increasing proportions of Cariflex IR 305 adversely affect strength properties and that this is carried through to even poorer properties after heat ageing.[14]

Industrial users have also reported that serious reductions in both tear strength and abrasion resistance occur above 25 parts replacement. However, with increased amounts of the synthetic polymer, compression set and resilience show some improvement.

7.2. Benefits of 'High' *cis*-Content
In contrast, the physical properties of synthetic 'high' *cis*-polyisoprene compare very favourably with those of high quality grades of natural rubber. Such a comparison of SMR CV and four commercially available polyisoprenes is shown in Table 8. In tests carried out at ambient

TABLE 7
COMPARISON OF NATURAL RUBBER WITH 'LOW' cis-POLYISOPRENE IN A CONVENTIONAL
BLACK-FILLED COMPOUND

SMR L	100	75	50	—
Cariflex IR 305	—	25	50	100
Compound viscosity ML 1 + 4, 100 °C	34	36	41	55
Hardness, IRHD	53	52	51	50
Modulus at 300%, MPa	4·4	4·2	3·9	3·2
Tensile strength, MPa	27·0	26·8	24·3	21·5
Elongation at break, %	650	670	680	710
Dunlop resilience at 20°C, %	92	92	93	93
Compression set 22 h 70°C, %	32	29	27	21
Aged properties (14 days at 70°C)				
Tensile strength, MPa	23·0	21·5	19·2	14·6
Elongation at break, %	570	615	600	580

Polymer, 100; Dutrex R, 3; stearic acid, 2; MT black, 30; sulphur, 2·5; CBS, 0·5;
TMTD, 0·1; Permanax ZA, 2; zinc oxide, 5. Cured to optimum at 140°C.

TABLE 8
COMPARISON OF PHYSICAL PROPERTIES OF 'HIGH' cis-POLYISOPRENES
(Test formulation to ASTM D3184-75-2A)

Physical properties	SMR CV	Natsyn 2200	Europrene IP80	Carom 2200	SKI 3S
Unaged					
Hardness, Shore A	59	62	57	57	57
Tensile strength, MPa	32·1	30·2	31·0	28·2	30·7
Modulus at 300%, MPa	14·0	13·3	11·7	11·2	12·5
Elongation at break, %	490	500	530	510	520
Tear strength, kN/m	116	103	107	93	111
Resilience at 23°C, %	83	79	78	74	78
Compression set, %	28·7	19·6	17·8	15·5	15·7
Abrasion loss, DIN	196	210	227	189	224
De Mattia cut growth					
2–4 mm, kc	4·0	6·0	9·0	22·0	12·0
4–8 mm, kc	5·0	7·0	19·0	38·0	18·0
Aged seven days at 70°C					
Hardness change, pts	0	−1	0	+2	0
Tensile strength change, %	−18	−13	−13	−4	−10
Elongation at break change, %	−16	−4	−9	−14	−6

Polymer, 100; zinc oxide, 5; stearic acid, 2; IRB 4, 35; TBBS, 0·7; sulphur, 2·25.
Note: Vulcanisates cured to Monsanto rheometer *t*′c (90) value at 150°C.

temperature, ultimate tensile properties compare closely for all the polymers but the synthetic 'high' cis-polyisoprenes generally have marginally lower tensile strengths and higher elongations at break. Lower modulus values are observed with all the synthetic polyisoprenes compared to natural rubber which may be ascribed mainly to the slower rate of strain-induced crystallisation. Relaxed moduli or load/deflection characteristics are, however, closely similar for both the natural and synthetic polymers.

These modulus differences are important since they are reflected in dynamic testing such as resilience and flexing. A low modulus polymer such as Carom 2200 will give vulcanisates with longer flex life and lower rebound resilience whereas those polymers with high moduli give shorter flex lives with higher rebound values. It is also of some significance that rebound resilience testing carried out at 100 °C shows results for natural rubber and Natsyn 2200 which invert those observed at ambient temperature, the synthetic polymer having higher rebound resilience.[15] This is directly related to the lower heat build-up characteristics observed for the synthetic polymer.[16] The Shore hardness values for Natsyn 2200 are somewhat higher than those shown either by natural rubber or the other 'high' cis-polyisoprenes. This is characteristic for this synthetic polymer over a wide range of formulations.

It should be noted that in this comparison, although the cure system was not specifically designed to give favourable compression set, the synthetic polymers all have lower set values than natural rubber. This again is a characteristic of synthetic 'high' cis-polyisoprenes.

Heat ageing characteristics of synthetic 'high' cis-polyisoprenes tend to be rather better than those of natural rubber. Changes in tensile strength and elongation at break after air oven ageing are less and this is illustrated in this study (Table 8).

7.3. Low Temperature Properties

In addition to the slower rate of crystallisation of synthetic polyisoprenes in the raw state relative to natural rubber, which was mentioned earlier, the effect persists in vulcanisates. It has been shown[17] by measuring the times ($t_{1/4}$) for 25 % stress relaxation at -26 °C of vulcanisates at 150 % extension that crystallisation is an order of magnitude slower for the synthetic rubber (Table 9).

Retardation of crystallisation increases with cure time in sulphur-containing vulcanisation due to formation of cyclic sulphides in the polyisoprene chain. Compounding techniques for inhibiting crystallisation have been published[18] and will apply equally to synthetic and natural

TABLE 9

CRYSTALLISATION QUARTER LIVES $(t_{1/4})$ AT $-26°C$ OF
VULCANISED NATURAL RUBBER AND SYNTHETIC POLYISOPRENE

	RSS1		Natsyn 2200	
Cure time at 140 °C, min	30	60	30	60
$t_{1/4}$, min	750	910	6 600	11 500

Rubber, 100; zinc oxide, 5; stearic acid, 1; Permanax WSP, 1;
MBT, 0·8; DPG, 0·1,; sulphur, 2.

polyisoprene. Where it is necessary to compound for very low temperature service conditions the use of synthetic rubber gives more flexibility by either reducing plasticiser levels or avoiding high sulphur systems which give poor ageing properties.

7.4. Anaerobic Ageing

Heat ageing in the presence of air is often used to categorise rubber compounds. Natural rubber and 'high' *cis*-polyisoprenes are not particularly resistant to ageing temperatures above 70 °C. For some applications, however, polyisoprene compounds may well exceed this temperature during their normal service life and often under dynamic conditions. This would be the case for compounds used in tyre carcass and for wire coat compounds where ingress of air or oxygen is limited.

It is interesting to compare the effect of ageing in air with ageing in an inert atmosphere such as nitrogen for both natural and synthetic polyisoprenes. Such a study has been carried out on a typical wire coat compound. For reasons of propriety the formulation cannot be fully divulged but it may be assumed to contain a relatively high loading of HAF black and a higher than normal sulphur level.

Air ageing and anaerobic ageing were carried out on vulcanisates under severe conditions and changes in ultimate strength properties and De Mattia cut growth performance were recorded (Table 10).

As expected, synthetic 'high' *cis*-polyisoprene gave original properties which showed increased hardness, marginally reduced tensile strength and greater elongation. Although all the compounds had relatively poor cut growth characteristics the synthetic polyisoprene-containing compounds performed marginally better.

After severe air ageing the substantial loss of strength properties was predictable and accompanied by increasing hardness. The compounds were very brittle and no De Mattia result was obtained.

TABLE 10
ANAEROBIC AGEING OF 'HIGH' *cis*-POLYISOPRENES

Composition			
SMR 10	100	50	—
Natsyn 2200	—	50	100
Physical properties, unaged			
Hardness, Shore A	70	71	74
Tensile strength, MPa	24·6	24·5	24·2
Elongation at break, %	500	510	520
De Mattia cut growth			
2–4 mm, cycles	1 500	1 500	2 000
4–8 mm, cycles	5 000	5 000	6 000
Aged seven days at 100°C in air			
Hardness, Shore A	79	80	82
Tensile strength, MPa	6·8	7·5	7·0
Elongation at break, %	40	40	40
De Mattia cut growth	Broke	Broke	Broke
Aged seven days at 100°C in nitrogen			
Hardness, Shore A	72	74	77
Tensile strength, MPa	8·6	10·2	11·1
Elongation at break, %	150	180	190
De Mattia cut growth			
2–4 mm, cycles	Broke at 1 500 cycles		3 000
4–8 mm, cycles			5 000

After anaerobic ageing ultimate strength properties were rather better than observed after air ageing and the compound containing all synthetic 'high' *cis*-polyisoprene not only gave better retention of tensile strength and elongation at break than those containing natural rubber but also gave De Mattia data which were very comparable with the original results.

This study was further extended to examine the effect of sulphur level on the compounds. The results obtained showed that for low and medium sulphur cure systems no particular benefit was obtained through using synthetic 'high' *cis*-polyisoprene. It would seem that the beneficial retention of dynamic properties given by synthetic 'high' *cis*-polyisoprene, and confirmed by other laboratory dynamic tests, is limited to compounds having high sulphur cure systems.

7.5. Compression Set
The superior compression set properties exhibited by synthetic polyisoprenes in conventional high sulphur/low accelerator vulcanisates was remarked upon in the foregoing section. It has been well established for natural rubber that substantial improvements in this property can be

achieved using efficient vulcanisation (EV) systems,[19] 'soluble' semi-EV systems[20] and peroxide cures. The improvements obtained by these methods are mainly due to the formation of thermally stable crosslinked networks which are intrinsically less prone to oxidation.

The compression set of synthetic polyisoprenes is also improved by these curing techniques and, moreover, they give substantially lower compression set than natural rubber in each of the vulcanisation systems.

TABLE 11

NATURAL AND SYNTHETIC POLYISOPRENES—COMPRESSION SET, %

	SMR CV	Natsyn 2200	Europrene IP80	Carom 2200	SKI 3S
Medium sulphur	17·5	11·8	11·6	11·5	12·0
Low sulphur	13·7	7·5	7·6	7·2	8·3
Sulphur-donor	11·0	6·1	6·9	6·0	7·5
Peroxide	13·8	9·7			

Base formulation—sulphur cures: Polymer, 100; zinc oxide, 5; stearic acid, 2; N-762 black, 75; Wingstay 100, 1.25.
Cure systems: Medium sulphur: CBS, 1.0; TMTD, 0.3; sulphur, 1.2. Low sulphur: TBUT, 0.6; ODEBS, 1.5; sulphur, 0.6. Sulphur-donor: MBTS, 2.0; TBBS, 1.5.
Formulation—peroxide cure: Polymer, 100; FEF black, 22; Flectol H, 1; Caloxol, 2; ZD_mC, 1.5; ZMBI, 0.5; Dicup 40, 6.
Test data: ASTM D395 Method B, 22 h at 70 °C.

This is illustrated in Table 11 for low sulphur and sulphur-donor EV systems, a semi-EV system and a peroxide cure. The peroxide-cured compounds would normally give lower compression set than the sulphur systems but the compound chosen in this case was also designed to give outstanding heat ageing and a compromise with low set was made. The practical implications of the lower compression set given by the synthetic rubbers are important since, not only can the very lowest compression set properties be obtained, but more cure system flexibility is offered. A less expensive system may be used to reach desired compression set properties and a better compromise can be reached between low compression set and good fatigue properties which are associated with higher sulphur levels.[21]

7.6. Creep

Creep may be defined as the progressive deformation of a vulcanisate under load with time.

7.6.1. Creep and Compression Set

Creep may appear initially to be directly related to compression set. A correlation is not obvious, however, even though low creep compounds generally exhibit good compression set. Creep occurs under continuous load, either under static or dynamic conditions, whereas compression set is a measure of the recovery after a load is removed. Also, both natural rubber and polychloroprene compounds have been successfully used for many years to support large heavy structures such as bridges or buildings where the creep property is extremely important since undue settlement of these structures with time must be avoided.

The low creep characteristics of mountings made with these rubbers are excellent provided that meticulous selection of ingredients and formulations is made.

7.6.2. Comparison of Creep Behaviour

Synthetic 'high' cis-polyisoprenes exhibit even better low creep properties than natural rubber in identical formulations and for this reason the synthetic polymers are finding increasing usage in high quality engineering applications.

A recent comparison of creep properties[22] was made on a specially constructed test jig which gave a 4:1 shear to compression ratio on a cylindrical sample. Measurements of primary creep, the increase in initial deformation as a function of log time (percentage per decade), were carried out on compounds specifically designed for low creep characteristics.[23] Both gum- and black-filled compounds were examined at ambient temperature (Table 12). The resulting data are interesting in that, not only are the differences between the natural and synthetic polyisoprenes

TABLE 12

CREEP CHARACTERISTICS OF NATURAL RUBBER AND SYNTHETIC POLYISO-PRENES (CREEP RATES IN PERCENTAGE PER DECADE UNDER A 4:1 SHEAR TO COMPRESSION RATIO AT 23 °C)

	Natural rubber SMR CV	Natsyn 2200	Europrene IP80	Carom 2200	SKI 3S
Gum	3·4	1·6	2·6	2·5	2·5
Black	5·1	3·0	4·2	3·8	3·8

Formulations: Gum: Polymer, 100; zinc oxide, 5; zinc 2-ethylhexanoate, 0.5; Flectol H, 1.0; Santoflex DPA, 1.0; TBUT, 0.6; Santocure MOR, 1.5; sulphur, 0.6. Black: As above plus N-762 black, 54.
Vulcanisation for each polymer $t'c$ (90) + 25%.

TABLE 13

CREEP CHARACTERISTICS OF NATURAL RUBBER, DEPROTEINISED
NATURAL RUBBER AND SYNTHETIC cis-POLYISOPRENE

	Creep rate (percentage per decade)
Natural rubber, SMR CV	3·2
Natural rubber, DPNR	2·4
Natsyn 2200	2·0

Compound: Polymer, 100; zinc oxide, 5; zinc 2-ethyl-hexanoate, 2; AgeRite resin D, 2; SRF black, 23; sulphur, 0.45; Santocure MOR, 2.2; TBUT, 0.88.

highlighted, but also a discrimination can be made between synthetic polyisoprenes from different sources. Natsyn 2200 would appear to be a more suitable material for low creep than some comparable rubbers and manufacturers who specialise in making low creep engineering components have indicated similar experiences, particularly for lower hardness compounds.

The reason for the lower creep exhibited by the synthetic polyisoprenes is undoubtedly due to a large extent to the absence of proteinaceous non-rubbers. It was demonstrated[23] that deproteinised natural rubber (DPNR) vulcanised using a soluble semi-EV system showed a significantly lower rate of stress-relaxation in extension in comparison to untreated natural rubber (SMR L). Further evidence of the deleterious effect of nitrogenous non-rubbers was given by creep measurements in 4:1 shear: compression (cited above) of SMR CV, enzyme deproteinised natural rubber and Natsyn 2200 in a black-loaded compound cured using an EV system. These results[24] given in Table 13 show that the creep rate of the DPNR is substantially reduced in comparison to the SMR CV and approaches the low creep rate exhibited by the synthetic polymer.

8. APPLICATIONS

8.1. General Utility
Synthetic 'high' cis-polyisoprene can be used for the whole range of products where natural rubber is the traditional choice. However, except for strategic reasons, for polymers produced in 'planned economy' countries, and for commercial reasons, in reality 'high' cis-polyisoprene does not find the same generalised utilisation as natural rubber. Present and planned capacities indicate that the ratio of 'high' cis-polyisoprene

consumption to that of natural rubber will continue to increase. In the interim the synthetic polymer is and will continue to be used for specific technical reasons of uniformity, purity and processing advantage as well as for final product quality and performance.

As for natural rubber the most important single outlet is in tyre production but benefits may be obtained in the manufacture of products as diverse as elastic bands, bridge bearings, conveyor belts and pharmaceutical goods. Certain specific attributes of synthetic polyisoprenes have made them attractive to manufacturers. For example, the lack of non-rubbers allows higher resistivity values to be reached in electrical applications and the low modulus provides easy extensibility and therefore comfort for items such as respirator masks and bathing hats.

The synthetic 'high' cis-polyisoprenes tend to be used most often in combination with the natural polymer since in this way the processing and uniformity properties of the synthetic polymer may be used to advantage whilst minimising compound changes. In addition the change from a natural rubber based formulation to a completely synthetic polymer formulation necessitates a different approach to mixing, handling and curing which has been discussed earlier and which may be unacceptable to some users.

8.2. Tyre Compounds

Since the polymers have been commercially available there has been considerable interest in both 'high' and 'low' cis-polyisoprenes for the modification of tyre compounds.

Since 'low' cis-polyisoprene is limited to approximately 20 % inclusion in natural rubber compounds due to the fall-off in physical properties, interest in this type of polymer has been specifically for processing reasons. Tyre performance has been shown to be unimpaired in tread formulations containing this level of both Cariflex IR 305 and IR 500 polymers.[25]

8.2.1. Tread Compounds

'High' cis-polyisoprenes may be used at much higher levels for tread compounds and even the complete replacement of natural rubber did not diminish abrasion resistance of tyres.[16] The same study highlighted a reduced heat build-up in truck tyre shoulders when using the synthetic polymer. Similar extensive investigations[26] in heavy duty tyre treads have confirmed that the use of 'high' cis-polyisoprene gave a cooler running tyre with no loss of abrasion resistance in comparison with all natural rubber tread compounds.

Substantial proportions of 'high' *cis*-polyisoprenes have been successfully incorporated into earthmover tyre treads to improve resistance to 'chunking'. Tear strength of the synthetic rubber has not been observed to be better than that of natural rubber, although very high strengths are obtainable by incorporation of minor amounts of reinforcing silica and a silane coupling agent. Therefore, it is not easy to see why use of the synthetic polymer is advantageous. A possible explanation could be that the tread of the tyre, when meeting stones or sharp asperities on rocks, can deform more easily due to the somewhat lower modulus characteristic, thereby obviating the tearing action.

The use of 'high' *cis*-polyisoprene is not limited to tread formulations and it is used to significantly greater advantage in other tyre components.

8.2.2. Breaker Compounds

With the advent of steel reinforced radial tyres, breaker compounds have become much more highly loaded with carbon blacks and 'green stocks' have consequently increased in viscosity. Such compounds require calendering and when based on natural rubber have a tendency to run very hot at this calendering stage. Blends containing 'high' *cis*-polyisoprene permit the use of high black-loadings and high viscosities but because of the flow characteristics of the synthetic polymer can be run cooler at the calender, thereby reducing scorch risk and improving dimensional stability of the breaker sheet. For natural rubber-based bead apex compounds the problem is similar but here the shaping process is by extrusion. Traditionally these compounds are heavily loaded and more recent demands by the tyre engineer for increasingly stiff compounds has necessitated incorporation of greater than 100 parts of highly reinforcing carbon blacks. Such compounds are difficult to mix, giving problems of both black incorporation and high power loads. The resultant high viscosities of mixed compound lead to some limitations on scorch and may be sufficiently high to exceed extruder drive-motor capacity.

'High' *cis*-polyisoprenes have been successfully used in this type of compound and even with compound Mooney viscosities up to MS (3 + 4 at 100 °C) of 110–120 have not caused undue problems at the extrusion stage.

8.2.3. Bead Wire Coat Compounds

Bead wire coat compounds are also generally tough and highly reinforced. Here the problem is excessive temperature build-up during processing which is sufficient to cause reversion of 'insoluble sulphur' to its normal form which is then capable of blooming to the surface of the compound giving subsequent adhesion problems. The use of 'high' *cis*-polyisoprenes, as for the breaker, permits cooler mixing and reduced extrusion temperatures, thus avoiding the problem.

8.2.4. Cord Adhesion

The adhesion of coating compounds to steel tyre cord is dependent upon intimate contact between the rubber compound and the actual wire surface. Since the steel wire used is not in single strands but in groups of several wires, to give additional strength, the minute interstices in the stranded wire may not be completely filled by natural rubber compound which retains a degree of retraction after the coating operation. A more intimate contact between the wire surface and the rubber compound is possible when using 'high' *cis*-polyisoprenes due to the easier flow characteristics of these polymers. Ultimate adhesion levels are not significantly higher than for natural rubber but variation of the adhesion level is markedly reduced.

8.2.5. Synthetic Polymer Characteristics

The characteristics of 'high' *cis*-polyisoprene compounds in anaerobic ageing conditions show improved aged property retention for certain types of formulations (see Section 7). Such compounds are particularly interesting in tyre carcass constructions where the ingress of air is minimal. Internal temperatures in heavy duty truck and bus tyres exceed the levels normally considered practical for either natural or synthetic polyisoprene. The retention of physical properties and the extended flex life of 'high' *cis*-polyisoprene under these anaerobic conditions can be beneficial to the tyre constructer.

The involvement and application of 'high' *cis*-polyisoprene for tyre construction may perhaps be best illustrated by the fact that an 'all synthetic' tyre is commercially feasible.[27]

This also shows that the lower compound 'green strength' associated with the synthetic polyisoprenes relative to natural rubber[28] is not an insurmountable problem. However, the incipient problem has been generally avoided by not making a total replacement of the natural rubber in tyre compounds which are subject to substantial deformation during building operations.

In addition to the specific features discussed, increased uniformity of compounds is probably the largest single attraction of 'high' *cis*-polyisoprene to the tyre manufacturer. Productivity gains during mixing and secondary processing are also achieved but are often difficult to quantify in financial terms.

8.3. Engineering Components

The particular properties of compression set and creep have already been discussed and the improvement in both these properties given by the replacement of natural rubber by 'high' cis-polyisoprene demonstrated.

The excellent resistance of 'high' cis-polyisoprenes to permanent set under load make them excellent materials for high quality engineering components such as springs, bearings, shock absorbers and bushes. Since the majority of these components are transfer- or injection-moulded the flow properties of synthetic polyisoprenes may be used to give the added advantage of rapid cycling and improved mould definition.

The load deflection characteristics of these types of component are often critical for their application. Natural rubber and the 'high' cis-polyisoprenes have load/deflection properties which are not dissimilar, however, the uniformity of the synthetic polymer is always greater than for the natural product and reproducibility of properties is assured. Since engineering components such as railway locomotive suspension units are often complex and extremely expensive to produce, this uniformity of product ensures a lower incidence of rejection.

Where high strength, good fatigue properties and low permanent set of components are required natural or synthetic polyisoprenes are almost universally used and, as engineers increasingly appreciate rubber as a material, higher demands for performance and longevity are made. Rubber engineering component manufacturers are appreciative of the extra benefits of 'high' cis-polyisoprene and it is now well established in the highest quality products.

8.4. Pharmaceutical Applications

Synthetic 'high' cis-polyisoprenes which are water-white in colour are stabilised with non-toxic antioxidants which conform to existing legislation on materials suitable for contact with food and drugs. Since they also have extremely low levels of impurities this makes them attractive polymers for pharmaceutical applications.

8.4.1. Baby Bottle Teats

Many products for surgical and pharmaceutical use, such as surgeons' gloves, are traditionally made from natural rubber latex and solid rubber is unlikely to be considered. However, one of the largest production areas where solid rubber products do compete strongly with latex products is in the manufacture of baby bottle teats. Moulded teats are generally made from first grade pale crepe which although having a near-white colour in the raw state gives vulcanisates which are brown. In spite of this coloration the final product has excellent transparency, an important factor for a product which relies upon a certain aesthetic quality at the point of sale.

It is possible with synthetic 'high' cis-polyisoprenes to produce baby teats which are water white in colour and which have a similar transparency to that found with pale crepe. It should be noted however that the final transparency given by a 'high' cis-polyisoprene depends upon its method of production. Manufacture of the synthetic polymer requires catalyst additions to effect polymerisation and these catalysts vary from one manufacturer to another. There is inevitably a certain small quantity of catalyst which remains in the finished polymer and this may give rise to vulcanisates from bottle teats which have a cloudy appearance although the colour is much lighter than that of teats produced from pale crepe.

Producers of bottle teats requiring improvement in colour of the finished product with minimal loss of transparency have found a suitable compromise in using compounds based upon synthetic 'high' cis-polyisoprene blended with a smaller amount of pale crepe which has the effect of 'solubilising' any residual catalyst in the synthetic polymer.

8.4.2. Other Pharmaceutical Products

'High' cis-polyisoprenes also find application in other pharmaceutical products. For catheters where extremely fine calendered film is used for the small inflators which prevent movement or dislodgement of this surgical insert, the synthetic polymer offers control of the tolerance of the rubber sheet used and can be calendered much thinner than natural rubber.

Bottle stoppers are other product areas where the synthetic polymers may be employed, both for reasons of colour and for the practical reason of resealability of stoppers used in phials which contain medicaments extracted by syringe.

The syringes themselves require plungers which exhibit not only purity but excellent sealing properties. The latter will depend upon the tolerances of moulding. The excellent mould definition and small controlled shrinkage experienced through using synthetic 'high' cis-polyisoprenes makes them ideal polymers for this application.

9.   MODIFIED POLYISOPRENES

9.1. Conflicting Trends

There are two conflicting trends affecting 'high' cis-polyisoprene production. The first, and most evident, is that the number of types offered by individual producers has diminished significantly and some offer only one grade. For instance, Goodyear, at one stage, had three basic polymers which differed mainly in the type of shortstop used. These were subdivided into particulate gel-free grades, lower Mooney viscosity grades and oil-extended versions. With refinements in equipment and polymerisation technique, the polymer is now free of particulate gel. The oil-extended types were withdrawn due to lack of demand. Now, apart from a very small demand for a lower Mooney rubber, one type has been retained which is universally accepted as having the best processing and cure characteristics and excellent storage stability.

The other trend is in the USSR, where in recent years, considerable efforts have been made to modify the basic polyisoprene with apparently one major objective in mind. In the course of discussions among rubber technologists, if synthetic polyisoprene is mentioned, this invariably provokes a comment on its low green strength characteristic. In the authors' opinion, the lower green strength of synthetic polyisoprene compared to natural rubber is not a barrier to its usage in the vast majority of situations and is largely a question of what one is used to. The efforts made in the USSR to increase the green strength of SKI 3 are no doubt strongly motivated by strategic and economic reasons and avoidance of dependence on natural rubber is an understandable goal.

9.2. 'No Gel' Polyisoprene
In the USSR a 'high' *cis*-polyisoprene had been manufactured[27] which was totally soluble in heptane at room temperature under static conditions compared to normal SKI 3 and which had a gel content of 20 %. The intention was to produce the most stereoregular polymer possible to obtain the ideal physical characteristics. The polymer had the usual content of *cis*-1,4 structure, somewhat higher molecular weight and was slightly less elastic in the raw state. It was claimed that it would permit improved processing and better physical properties but as no details were given, we assume that little progress was made by eliminating gel and that green strength was unaffected.

9.3. Polyisoprenes with Functional Groups
SKI 3 had been modified to introduce anhydride and unspecified amine groups[27] to give polymers which showed increased green strength in black-filled compounds but not in gum stocks. The anhydride-modified rubber showed the larger increase but also had a high gel content in the raw state. Treatment of black compounds of these two polymers with polar and non-polar solvents demonstrated that the amine-modified type interacted only

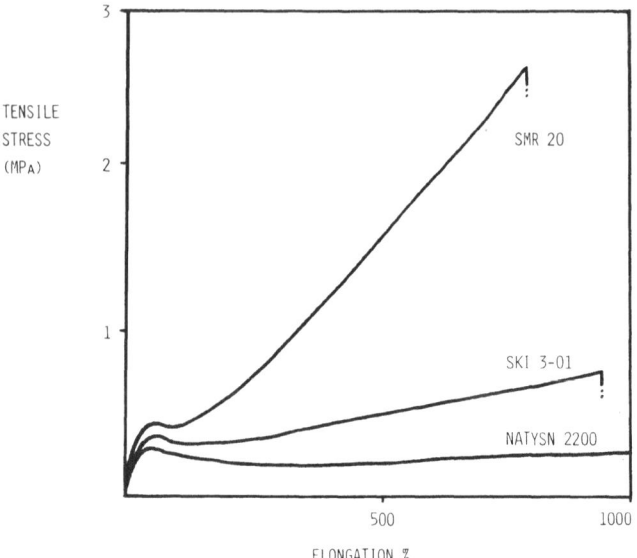

FIG. 9.   Stress–strain characteristics of unvulcanised compounds (polymer, 100;
zinc oxide, 5; stearic acid, 2; N-330 black, 50; aromatic oil, 3).

with the black while the anhydride groups in the other polymer interacted with themselves as well as with the black surface.

Hydroxyl groups have also been introduced into SKI 3, which gave a modest increase in green strength, but required addition of an isocyanate in the compound to give a large effect. Inferior tensile, tear and dynamic properties of the isocyanate-treated compound resulted and it is likely that processing problems would be encountered.

Cyclocarbonate and bromine groups have been introduced into SKI 3 experimentally[30] and with added benzylamine in the compounds, increased green strength was obtained but again at the expense of tensile properties, although resilience was improved.

cis-Polyisoprene modified with 4-nitrosodiphenylamine has been produced commercially, SKI 3–01,[31] and is claimed to give green strengths approaching those of natural rubber in a tyre carcass compound. The physical properties of the vulcanisates were closely similar to those of the unmodified SKI 3, with improvements in both resilience and flex life.

A comparison made in our laboratories using an HAF black-loaded compound showed that although the green strength was improved over an unmodified polyisoprene it was still well below that of natural rubber (Fig.

9). The mill handling behaviour of the SKI 3–01 compound was not attractive, in that it was akin to natural rubber in respect of nerve and exhibited stickiness. This indicates that any gain in green strength is likely to be offset by losses in easy processing in a factory situation.

9.4. Use of Promoters

It has been established for some years that promoters, such as N-(2-methyl-2-nitropropyl)-4-nitrosoaniline (Nitrol) offer the possibility to increase the green strength of cis-polyisoprenes to levels in excess of 1 MPa at a breaking elongation of about 900% in a similar compound to that described in Fig. 9. This requires an extra step in the mixing cycle and the compound is distinctly less easy to process as a result of the increased polymer/black interaction. In our experience, the technique is seldom used to solve a green strength problem, probably because the easy processing of the synthetic polyisoprene is lost and it may be that the quest for a high green strength synthetic polyisoprene is futile for the same reason.

ACKNOWLEDGEMENT

The authors wish to acknowledge the contributions of many colleagues within the Goodyear Organisation over a long period of time.

NOTE

While the information herein is believed to be reliable and correct, nothing herein is intended and should not be construed, as a representation or warranty, express or implied, as to results obtained or to be obtained by others who may make use of this information or with respect to the absence, existence or validity of patent rights, if any, of others involving any composition or process herein referred to; or an inducement or recommendation for the violation of any such patent rights; and responsibility and liability therefore is disclaimed.

REFERENCES

1. SCHOENBERG, E., MARSH, M. A., WALTERS, S. J. and SALTMAN, W. M. *Rubber Chem. Technol.*, **52**, 1979, 526.
2. TIDD, B. K. Malaysian Rubber Producers' Research Assn., 1976, private communication.
3. SEKHAR, B. C., *J. Polym. Sci.*, **48**, 1960, 133.
4. SEKHAR, B. C., British Patent 965,757, 1964.
5. ELLIOT, D. J. *Developments in Rubber Technology—1*, Applied Science Publishers, London, 1979, 4.
6. CHUBB, S. C. G. and DOYLE, G. M. *J. IRI*, **4**, 1970; 30.

7. BRUZZONE, M. *et al. Proc. 4th International Synthetic Rubber Symposium*, *London*, 1969, 83.
8. BRISTOW, G. M. *J. Polym. Sci.*, **62**, 1962, S168.
9. ANON. *NR Technol.*, **10**, 1979, 41.
10. BRISTOW, G. M., CUNNEEN, J. I. and MULLINS, L., *NR Technol.*, **4**, 1973, 16.
11. BAKER, H. C. and GREENSMITH, H. W. *Trans. I.R.I.*, **42**, 1966, T194.
12. WHEELANS, M. A., *NR Technol.*, **9**, 1978, 71.
13. HANNSGEN, F. W. *Rubber and Plastics Age*, **42**, 1961, 166.
14. LIM, S. K. and WATSON, A. A. Malaysian Rubber Producers' Research Assn., 1970, unpublished data.
15. BRISTOW, G. M., CUNNEEN, J. I. and MULLINS, L. *International Symposium on Isoprene Rubber*, Moscow, 1972.
16. ALARI, G. and BRUZZONE, M. *International Symposium on Isoprene Rubber*, *Moscow*, 1972.
17. BRISTOW, G. M. Malaysian Rubber Producers' Research Assn., 1968, private communication.
18. ANON. *NR Technol.*, **1**, 1970 Part 1, 4.
19. SKINNER, T. D. and WATSON, A. A. *Rubber Age*, **99**(11), 1967, 76.
20. ELLIOT, D. J., SKINNER, T. D. and SMITH, J. F. *International Rubber Conference, Moscow*, 1969.
21. AYERST, R. C., LLOYD, D. G. and RODGER, E. R. *DKG Meeting*, Weisbaden, 1971.
22. STIWALD, E. C. Goodyear Tire and Rubber Co. Akron, 1979, unpublished data.
23. GREGORY, M. J. *NR Technol.*, **8**, 1977, 1.
24. FATH, M. A., Goodyear Tire and Rubber Co., Akron, 1973, unpublished data.
25. KLEEMAN, W. and ERBEN, G. *International Symposium on Isoprene Rubber*, *Moscow*, 1972.
26. SARBACH, D. V., HALLMAN, R. W. and SUDEKUM, J., *Rubber Age*, **91**(11), 1969, 53.
27. MASAGUTOVA, L. V., POLUKETOVA, L. E., TROITSKAYA, N. I., SAPRONOV, V. A., EVSTRATOV, V. F. and ZHAKOVA, V. G. *International Rubber Conference, Venice*, 1979.
28. KNILL, R. B., SHEPHERD, D. J., URBON, J. P. and ENDTER, N. G., *International Rubber Conference, Kuala Lumpur*, 1975.
29. TODD, R. V., *Rubber and Plastics Age*, **49**, 1969, 921.
30. TROSTYANSKAYA, I. I., BOLKHOVETS, B. M., SMIRNOV, V. P., RAPPOPORT, L. Y., KOVALEV, Y. F. and PETROV, G. N. *Int. Polym. Sci. Technol.*, **6**(II), 1979, T/57.
31. KOGAN, L. M., SMIRNOV, V. P., KOVALEV, N. F. and KROL, V. A. *Int. Polym. Sci. Technol.*, **6**(5), 1979, T/20.

APPENDIX: LIST OF REGISTERED TRADE NAMES AND CHEMICAL ABBREVIATIONS

AgeRite: R. T. Vanderbilt Co. Inc., Norwalk, Conn., USA.
Ameripol: B. F. Goodrich Chemical Group, Cleveland, Ohio, USA.

Caloxol: John and E. Sturge Ltd, Birmingham, UK.
Cariflex: Shell Chemical Co., Houston, Texas, USA.
Carom: Chemimportexport, Bucharest, Rumania.
CBS, N-cyclohexyl benzthiazole-2-sulphenamide.
Dicup: Hercules Inc. Wilmington, Delaware, USA.
DPG, diphenylguanidine.
Dutrex: Shell Chemical Co., Houston, Texas, USA.
Europrene: ANIC S.p.A., Milan, Italy.
Flectol: Monsanto Co., Akron, Ohio, USA.
MBT, 2-mercaptobenzthiazole.
MBTS, di-(benzthiazole-2-yl)disulphide.
Monsanto rheometer, Monsanto Co., Akron, Ohio, USA.
Natsyn: The Goodyear Tire & Rubber Co., Akron, Ohio, USA.
Nipol: Nippon Zeon Co. Ltd, Tokyo, Japan.
Nitrol: Monsanto Co., Akron, Ohio, USA.
ODEBS, N-oxydiethylene benzthiazole-2-sulphenamide.
Permanax: Vulnax International Ltd, Blackley, Manchester, UK.
Plasti-corder: Brabender OHG Duisberg, West Germany.
Processability tester, Monsanto Co., Akron, Ohio, USA.
Santocure NS (N-t-butyl benzthiazole-2-sulphenamide): Monsanto Co.,
   Akron, Ohio, USA.
Santoflex: Monsanto Co., Akron, Ohio, USA.
SKI produced in USSR, v/o Raznoimport, Moscow, USSR.
TBBS, N-t-butyl benzthiazole-2-sulphenamide.
TBUT, tetrabutylthiuram disulphide.
TMTD, tetramethylthiuram disulphide.
Wallace rapid plastimeter, Wallace Instruments Ltd, Croydon, UK.
Wingstay: The Goodyear Tire & Rubber Co., Akron, Ohio, USA.
ZD_mC, zinc dimethyldithiocarbamate.
ZMBI, zinc 2-mercaptobenzimidazole.

INDEX